Stochastic Causality

Stochastic Causality

edited by
Maria Carla Galavotti
Patrick Suppes
Domenico Costantini

CSLI
Publications
Center for the Study of
Language and Information
Stanford, California

Library of Congres Cataloging-in-Publication Data

Stochastic causality / edited by Maria Carla Galavotti, Patrick Suppes,
Domenico Costantini.
   p. cm. -- (CSLI lecture notes ; no. 131)
   Includes bibliographical references and indexes.
   ISBN 1-57586-321-9 (cloth : alk. paper) -- ISBN 1-57586-322-7 (pbk.
: alk. paper)
   1. Causation. 2. Probabilities.   I. Galavotti, Maria Carla. II.
Suppes, Patrick, 1922-   . III. Costantini, Domenico. IV. Series.
   BD541 .S68 2001
   122--dc21
                          2001047485

∞ The acid-free paper used in this book meets the minimum requirements of the
American National Standard for Information Sciences – Permanence of Paper for
Printed Library Materials, ansi z39.48-1984.

Please visit our web site at
http://cslipublications.stanford.edu/
for comments on this and other titles, as well as for changes
and corrections by the authors, editors and publisher.

Contents

Contributors

NANCY CARTWRIGHT: Department of Philosophy, Logic and Scientific Method, The London School of Economics & Political Science, Houghton Street, London WC2A 2AE and Philosophy Department, University of California, San Diego, 9500 Gilman Drive, La Jolla CA 92093-0119 (January-April, yearly).
PHILCENT@lse.ac.uk and ncartwright@popmail.ucsd.edu

DOMENICO COSTANTINI: Department of Statistics, University of Bologna, via Belle Arti 41, 40126 Bologna, Italy.
costanti@stat.unibo.it

MARIA CARLA GALAVOTTI: Department of Philosophy, University of Bologna, via Zamboni 38, 40126 Bologna, Italy.
galavott@philo.unibo.it

UBALDO GARIBALDI: CFSBT-CNR, c/o Department of Physics, University of Genoa, via Dodecaneso 33, 16146 Genoa, Italy.
garibaldi@fisica.unige.it

GIANCARLO GHIRARDI: Department of Theoretical Physics, University of Trieste, Strada Costiera 11, Miramare Grignano, 34014 Trieste, Italy.
ghirardi@trieste.infn.it

PAUL W. HOLLAND: Educational Testing Service, Rosedale Road, Mailstop 16-T, Princeton, New Jersey 08541.
pholland@ets.org

GÜROL IRZIK: Philosophy Department, Boğaziçi University, 80815 Bebek Istanbul, Turkey.
irzik@boun.edu.tr

DAVID PAPINEAU: Department of Philosophy, King's College London, Strand, London WC2R 2LS, UK.
david.papineau@kcl.ac.uk

MICHEL PATY: Equipe REHSEIS (UMR 7596), CNRS et Universite Paris 7 Denis Diderot, Centre Javelot, 2, Place Jussieu, F-75251 Paris Cedex 05, France.
paty@paris7.jussieu.fr

HUW PRICE: Department of Philosophy, University of Edinburgh, David Hume Tower, George Square, Edinburgh EH8 9JX, Scotland.
Huw.Price@ed.ac.uk

ROBERTO SCAZZIERI: Department of Economics, University of Bologna, Piazza Scaravilli, 2, 40126 Bologna, Italy.
scazzieri@economia.unibo.it

WOLFGANG SPOHN: Fachbereich Philosophie, Universität Konstanz, Fach D 21, 78457 Konstanz, Germany.
wolfgang.spohn@uni-konstanz.de

PATRICK SUPPES: Center for the Study of Language and Information, Stanford University, Stanford, CA 94305-4115.
suppes@csli.stanford.edu

ALESSANDRO VERCELLI: Department of Economics, University of Siena, Piazza S. Francesco, 7-53100 Siena, Italy.
vercelli@unisi.it

ATTILIO WEDLIN: Department of Economics and Statistics, University of Trieste, Piazzale Europa 1, 34127 Trieste Italy.
attilio.wedlin@econ.univ.trieste.it

JAMES WOODWARD: Division of the Humanities and Social Sciences 101-40, 209 Dabney Hall, California Institute of Technology, Pasadena, CA 91125.
jfw@hss.caltech.edu

Preface and Acknowledgements

The present volume is meant to contribute to the extensive current literature on the nature of causality, and in particular, probabilistic causality in a temporal context, whence the title *Stochastic Causality*. Starting around the beginning of the last century, causality was in disgrace for a long period, after the advent of the new physics had cast doubt on the deterministic paradigm. These developments seemed to suggest good reasons for banning causality from the realm of science as a "relic of a bygone age", as Bertrand Russell called it in 1913 in "On the Notion of Cause". Causality, however, has not only survived in the work of many scientists in various fields, but has been revived within philosophy as well. Its resurgence is largely due to its reformulation in probabilistic terms, which started around the middle of the last century.

Once causality is defined in probabilistic terms, i. e., the notion of causal link is extended from constant to probable conjunction, the problem arises immediately of how to keep causal relations separate from spurious correlations, a problem which has a long tradition in the statistical literature. In the hands of philosophers, this problem has raised a whole array of issues, connected to the Humean or non-Humean character of probabilistic causation, and the possibility or impossibility of reducing causal links to probabilities. A further problem generated by the probabilistic approach is the relationship between token-causality, or causality referred to single events, and property causality, or causality referred to populations of events. Other problems pertaining to the notion of causality have received attention within a probabilistic framework, such as the relation between causality and counterfactuals, as well as that between causality and explanation.

A conception of causality that has become increasingly popular in the last decades is in terms of manipulability. The underlying idea is that a property can be taken as causal, when by changing it, it is possible to induce changes in another property. A long tradition in econometrics has associated causality with manipulability and has defined it with reference to models. The specification of causal models in this context raises special problems, which are presently the object of much attention on the part of statisticians and econometricians.

While in its traditional, deterministic formulation causality used to be primarily a philosophical topic, in its probabilistic version it is a rich concept, of interest to philosophers, statisticians and scientists. Being a genuinely interdisciplinary notion, stochastic causality presents problems that can be tackled through the combined efforts of people working in different fields. A central interest of this collection lies in the fact that its contributors come from various disciplines—economics, philosophy, physics and statistics. The diversity of their backgrounds is reflected in the variety of their analyses of causal concepts.

The first three articles (Galavotti, Papineau, Woodward) focus on probabilistic causality in the recent philosophical tradition, with a special emphasis on manipulation by Galavotti, a focus on metaphysics and methodology by Papineau, and consideration of the place of counterfactuals by Woodward. Galavotti is concerned to integrate Salmon's mechanics-based theory of causality with the manipulative view of Price and Menzies. Papineau returns to the defense of the reduction of causes to probabilities, by arguing that the critics of this reduction have muddled metaphysics and methodology. Woodward argues that, contrary to the views of many philosophers, causal modeling, which uses, in addition to probabilities, either structural equations or directed graphs, is closer to the counterfactual approach than to probabilistic accounts of causality. In the fourth article Cartwright examines the issue and efficacy of modularity. Her focus, as in her recent book, is to celebrate the particular over the universal in causality, as in other matters. Next, Irzik examines three Humean dogmas of causation. He is especially concerned to challenge the epistemological dogma that our knowledge of causes is always indirect, because it is based on observed regularities. In the sixth article, Price makes a case for pragmatism about causation in the special sciences. In this context by *pragmatism* he means the study of the scientific practice in a given discipline of how causal notions are used.

The seventh and eighth articles are by two economists. Scazzieri discusses various aspects of causality in the context of economic choice. His emphasis is on using bounded rationality as the appropriate choice model, and to concentrate on such features of choice behavior as pattern recognition and induction by analogy. Vercelli looks at causality and uncertainty from a Keynesian viewpoint, reaching back to Keynes' early treatise on probability.

He emphasizes Keynes' anticipation of several concepts that only later received prominence, such as the tension between *causa cognoscendi* and *causa essendi*.

The next three articles are broadly statistical in character. Spohn takes up the currently fashionable question whether or not Bayesian nets are all there is to causal dependence. Holland analyzes the often-discussed causal interpretation of regression coefficients, especially important because of the extensive use of linear regression models in many different scientific disciplines. Wedlin looks at partial exchangeability, which generalizes simple exchangeability. He extends de Finetti's original idea of partial exchangeability to hierarchical causal models.

The last four articles are centered on problems of causality in physics. Suppes focuses on weak and strong reversibility in classical mechanics as well as in stochastic processes of several types. Costantini and Garibaldi give a detailed qualitative characterization of creation and destruction propensities, and then develop the associated probabilistic quantitative formulation. In doing so, they bring out the sharp conceptual difference between destruction and creation of particles in statistical mechanics. Paty explores, with attention to historical developments, the concept of state and the meaning of probability in quantum mechanics. His emphasis is on epistemological analysis, and he argues, e.g., contrary to the standard view, that the state function represents directly the physical system, not our knowledge of it. Ghirardi examines the relation between stochastic dynamical reduction and causality in the well-known GRW model of quantum phenomena, of which he is a principal proponent. The focus of philosophical concern is how to analyze causal relations between events, given the necessary nonlocal nature of the theory.

The fifteen articles collected in this volume were originally presented at two conferences that were jointly organized by the editors. The first meeting took place at Ventura Hall in Stanford on April 16-17, 1998. Eight papers were read in this order by James Woodward, Yair Guttmann, Domenico Costantini, David Freedman, Paul Holland, Brian Skyrms, Maria Carla Galavotti and Patrick Suppes. The second was held at the Conference Centre of the University of Bologna in Bertinoro, on September 20-22, 1999. The meeting hosted fourteen papers, delivered in this order by Patrick Suppes, Huw Price, Gurol Irzik, Matthias Hild, Attilio Wedlin, GianCarlo Ghirardi, Michel Paty, Yair Guttmann, David Papineau, Nancy Catrwright, Alessandro Vercelli, Roberto Scazzieri, Wolfgang Spohn, and jointly, Domenico Costantini and Ubaldo Garibaldi. Both conferences devoted ample space to discussion and gave the participants the opportunity for extended exchange of ideas.

In publishing the proceedings of the two conferences, we wish to thank all those who took part in them, both as speakers and discussants, and especially those who agreed to contribute to the proceedings. Thanks are also due to those who supported in various ways our initiative: the Centre of Interdisciplinary Research "Federigo Enriques" and the Departments of Philosophy, Statistics and Economics of the University of Bologna, who supported the Bertinoro meeting, and the Center for the Study of Language and Information of Stanford University for support of the first meeting. Finally, thanks to Ann Gunderson for extensive editorial work in preparing the volume for publication.

Maria Carla Galavotti
Patrick Suppes
Domenico Costantini

1

Causality, Mechanisms and Manipulation

MARIA CARLA GALAVOTTI

University of Bologna

1 Introduction

This paper suggests an integration of Wesley Salmon's mechanistic theory
of causality with a manipulative account of causation of the kind that has
been recently defended by Huw Price and Peter Menzies. Firstly, Salmon's
view of causality is outlined, and the main issues of the debate around it are
recollected. Secondly, the manipulative view of causality is sketched and
the possibility of its integration with Salmon's theory is considered for the
purpose of coping with some of the problems raised by its critics.

2 Two Kinds of Probabilistic Causality

After having been for centuries an essential component of the mechanistic
picture of the world, where it is strictly connected to explanation, causality
underwent a crisis after the deterministic paradigm was put in doubt by the
new physics. Not surprisingly, causality plays a minor role within Hempel's
model of explanation, which has long been considered the official theory of
scientific explanation developed by philosophers of science, ever since it
was put forward in the early forties.

The resurgence of causality after a period of disgrace is linked to its
probabilistic interpretation. The literature on probabilistic causality started

Stochastic Causality.
Maria Carla Galavotti, Patrick Suppes and Domenico Costantini (eds.).
Copyright © 2001, CSLI Publications.

with a few important, albeit isolated, works, like Hans Reichenbach's *The Direction of Time* (1956), Irving John Good's "A Causal Calculus" (1961-62) and Patrick Suppes' *A Probabilistic Theory of Causality* (1970), and has been constantly growing since. The probabilistic treatment of causality leads immediately to a distinction between causal talk referring to population variables, or "property causality", and causality between single events, often called "token" or "aleatory" causality.

The difference between the two was recognized at the outset by Good, who distinguished between "the tendency of F to cause E" and "the degree to which F caused E" (Good 1961, p. 307). For Good, these are different notions, resting on different probabilistic assumptions and requiring different probability measures.

In his monograph on probabilistic causality, Suppes proposed two definitions, one in terms of events and another in terms of statistical variables. However, it is clear that he was thinking of kinds of events, or property causality. As a matter of fact, most of the criticisms and counterexamples raised against Suppes' theory of causality are (explicitly or implicitly) centred on the lack of distinction between two different levels of analysis[1]. In reply to his critics, Suppes came to an explicit recognition of two different kinds of causality. So much granted, he favors a theory of probabilistic causality referred to kinds of events, claiming that what matters in science is the behaviour of populations, more than that of single events (see Suppes 1984).

The tension between property and token causality reflects the tension between prediction and explanation in probabilistic contexts. Property causality has predictive power, but differs from predictability, since one can usually make predictions based on mere statistical correlations. On the contrary, single events are often unpredictable, and can only be explained after they occur. The information at our disposal is generally based on knowledge regarding statistical relationships, and these do not bear directly on explanation. This is why Suppes' theory is not intrinsically linked to explanation, and Good treats the notion of "explicativity" independently from that of causality.

3 Salmon's Mechanistic View

A major attempt to put together explanation and probabilistic causality has been made by Salmon, who grounds the notion of explanation on that of causality, in the conviction that genuine explanation is causal in character,

---

[1]A case in point is D. Rosen's example of the golf ball falling into the hole after hitting the branch of a tree. The example is discussed in Suppes (1970) and (1984).

and that explanatory power rests on causal links. Salmon's aim is to revive mechanical explanation in a probabilistic framework.

His first attempt to do this is contained in *Statistical Explanation* (1970). Building on the notion of statistical relevance, he put forward a model of explanation meant as an alternative to Hempel's approach. This is the Statistical-Relevance model of explanation, according to which an event is explained by showing what factors are statistically relevant to its occurrence. To accomplish this aim, the reference class to which the event belongs is gradually restricted, to make it include only the relevant factors. In this process, irrelevant factors are "screened off" by relevant ones until a homogeneous reference class is obtained. The fact to be explained is eventually located in a network of statistical relations holding between the properties that are relevant to its occurrence. One can say that the *explanandum* event is explained when its place within such a network is specified. In other words, explanation associates the event to be explained with a probability distribution.

When referred to a homogeneous reference class, the probability distribution associated with the *explanandum* reflects the most complete and detailed information attainable. Homogeneity of the reference class then becomes the main requirement of explanation conceived in this way. However, given the obvious difficulties connected with the notion of "objective homogeneity", Salmon takes it as characterizing the ideal case to which explanations should tend. In those cases in which objectively homogeneous reference classes are not obtainable, epistemically homogeneous reference classes are adopted.

Clearly, the S-R model conveys information on the statistical relationships holding among the variables of a population. But for someone like Salmon, who thinks that explanatory information has to be causal, statistical correlations themselves call for an explanation. In view of this, Salmon works out a second level of explanation with respect to which the S-R model becomes a first step, also called "S-R basis". Explanation of the second level is causal and includes a probabilistic theory of causality.

To substantiate probabilistic causality, Salmon adopts the notion of "causal process", defined as a spatio-temporal continuous entity having "the capacity... to transmit information, structure and causal influence" (Salmon 1994, p. 303). Causal processes are responsible for causal propagation and provide the links between causes and effects. They can be thought of as forming a net, whose knots represent interactions between processes. When processes are modified in such interactions, causal production takes place. The changes thus obtained persist after the interaction and are propagated by the processes. The basic idea that enters in the definition of causal nets is the "principle of the common cause" which Salmon borrows from Reichen-

bach with some modifications, mainly due to the fact that in a causal interaction the common cause does not "absorb" the probability of the joined
effects. (Instead of the equality $P(A \cdot B \mid C) = P(A \mid C) \cdot P(B \mid C)$, we have
the inequality $P(A \cdot B \mid C) > P(A \mid C) \cdot P(B \mid C)$.)

Causal processes and interactions, however, cannot be defined only in
terms of relationships among probability values. According to Salmon,
speech in terms of causal processes and interactions makes reference to
physical properties, which form the mechanisms responsible for the occurrence of phenomena. Causal explanation obtains when phenomena are
located at some point within the net of causal processes, and it tells us how
such mechanisms work. Salmon's theory of causal explanation is contained
in the volume *Statistical Explanation and the Causal Structure of the World*
(1984), which has provoked extensive debate.

This revolves partly around the notion of "causal process", beset with
various difficulties pointed out, among others, by P. Kitcher and P. Dowe.
Their main criticism regards the appeal to counterfactuals contained in
Salmon's first formulation of processes in terms of "mark transmission".
After having amended it a few times, Salmon proposed a formulation in
terms of "invariant quantities" (Salmon 1994) and, more recently, one in
terms of "conserved quantity" (Salmon 1997). The issue is under debate.

A further topic of concern is the relationship between the two levels of
explanation envisaged by Salmon. After the publication of (1984), in a series of papers now included in the collection *Causality and Explanation*
(1998), Salmon gradually reached the conclusion that the two levels of
analysis, namely the S-R basis (or "statistical causality") and causal talk in
terms of processes and interactions (or "aleatory causality") cannot be
brought together. Let me briefly recollect how Salmon reached this conclusion.

Around 1990 Salmon thought that only aleatory causality could give
"an adequate understanding of causality" (Salmon 1998, p. 207). This conviction has been challenged by C. Hitchcock on the claim that, while envis-
aging a geometrical network of processes and interactions, Salmon's theory
does not contain any hint as to what properties should be taken as explanatory. In contrast, Hitchcock favours a stronger notion of explanation, and
endorses Woodward's claim that explanation answers "what-if-things-had-
been-different" questions. Following this path, he is led to the conclusion
that "a successful account of explanation had better make the relation of
explanatory relevance look roughly like that of counterfactual dependence"
(Hitchcock 1995, p. 311), and that "our demand that explanations provide
relevant information requires something stronger—that we be told *which*
earlier properties the properties specified in the *explanandum* depend upon"
(*ibidem*). On this basis, Hitchcock rejects Salmon's distinction between two

levels of explanation and claims that explanatory relevance should rest on the counterfactual information given by statistical correlations.

In response to Hitchcock, Salmon has reconsidered the link between the two levels of explanation, coming to the conclusion that "(1) statistical relevance relations, in the absence of connecting causal processes, lack explanatory import and (2) that connecting causal processes, in the absence of statistical relevance relations, also lack explanatory import... Both are indispensable" (Salmon 1997, p. 476). The two levels of explanation are thereby strictly connected. The causal model in terms of processes and mechanisms is accordingly seen as an essentially geometrical model, that needs to be implemented with information on statistical relevance relations in order to allow recognition of those properties which are pertinent to given outcomes.

At the same time, Salmon concedes that counterfactual considerations have a role to play within explanation and reaffirms a close connection between statistical relevance and counterfactual information. After banning counterfactuals from the definition of causal processes, he admits them at the explanatory level, claiming that such counterfactuals are "relatively unproblematic", being supported by well-established statements of statistical correlations, based on observed frequencies. The question of a more detailed definition is not addressed.

As a further concession to his critics, Salmon recognizes that pragmatical considerations have a role in determining what is to be taken as explanatorily relevant. This move is made necessary by the fact that causal talk in terms of processes can be conducted at different levels of analysis. Sometimes facts can be accounted for at the most detailed level allowed by scientific theories, but in most cases phenomena are analysed at different levels of abstraction, as determined by the context.

The far from straightforward story of the development of Salmon's theory testifies the difficulties one encounters when trying to combine property and token causality within a unified theory of (probabilistic) explanation and prediction. In Salmon's view the asymmetry of causation is made to depend ultimately on spatio-temporal continuous processes, and such a notion is much too strong to guarantee the model-wide applicability. The same holds for the homogeneity requirement, which usually does not apply to the social sciences. This is stressed by J. Woodward, who observes that in this field variables like class, religion, education and the like are taken as explanatory even though finer partitions would be possible (Woodward 1989). Though many agree that Salmon's theory captures our intuitions about explanation and causality, as well as important features of explanation in classical and statistical physics, its application to other fields encounters major difficulties.

In (Galavotti 1999) I suggested a possible integration between Salmon's model and a weaker notion of causality, like the manipulative view that has been worked out in some detail by the literature on econometric models. A similar notion of causality has recently been the object of a perceptive treatment by the epistemologist H. Price, partly in collaboration with P. Menzies.

4 The Manipulative View

The central idea of Price's approach is that, instead of being taken as a property of the world, causality can be related to the agent's perspective, in view of the fact that we acquire the notion of causation through our experience as agents. "Roughly, to think of A as a cause of B is to think of A as a potential *means* for achieving (or making more likely) B as an *end*" (Price 1992a, p. 514). The main advantage of this view is that it embodies in a natural way the asymmetry of causation, which derives from that characterizing the means-end relationship. In Prices' words: "causes are potential means, on this view, and effects their potential ends. Causal asymmetry originates in our experience of doing one thing to achieve another; in the fact that in the circumstances in which this is possible, we cannot reverse the order of things, bringing about the second state of affairs in order to achieve the first" (p. 515). Price ascribes the paternity of this conception to F. P. Ramsey, who thought that, like other notions, including probability, laws and chance, causation is ultimately rooted in the perspective with which an agent looks at the world (see Price 1992b).

The general formulation of the agency theory of causation given by Price and Menzies is probabilistic, but they add that "it seems reasonable to expect that an indeterministic theory will incorporate the deterministic notion, as a special or limiting case" (Menzies and Price 1993, p. 189). The means-end relation on which the theory is grounded is characterized in terms of what they call "agent probabilities". The latter "are to be thought of as conditional probabilities, assessed from an agent's perspective under the supposition that the antecedent condition is realized *ab initio*, as a free act of the agent concerned. Thus the agent probability that one should ascribe to B conditional on A...is the probability that B would hold were one to choose to realize A" (p. 190).

Though this position is admittedly anthropocentric, and takes causality to be context-dependent, according to its proponents it has decisive advantages over more traditional approaches. The fact that it accounts in a natural way for causal asymmetry, which is embedded in the notion of agency, has two important consequences. First, in this perspective the arrow of causation does not need to be based on the arrow of time. For Price this is a major advantage, as it allows for contemporary and even backwards causation, which plays a decisive role in his interpretation of microphysics (see Price 1996).

Furthermore, the agency view offers a solution to the problem of spurious causation, because it assigns statistical correlations an asymmetrical interpretation in the light of the agent's (probabilistic) judgment in terms of the means-end relation. So, according to Menzies and Price, "Agent probabilities fail to generate spurious correlations because they abstract away from the evidential import of an event, in effect by creating for it an independent causal history. For example, in enquiring whether one's manipulation of an effect B would affect the probability of its normal cause A, one imagines a new history for B, a history that would ultimately originate in one's own decision, freely made. And one thereby deprives B of its usual evidential bearing on A. The same is true, *mutatis mutandis*, for the situation in which one manipulates one effect of a common cause to see whether it makes a difference to the probability of another effect" (Menzies and Price 1993, p. 191). From the point of view of the agent, correlations due to common causes do not necessarily translate into probabilistic dependencies. Consequently, from the point of view of the agent there is no divergence between probabilistic and causal relevance, they just coincide.

An additional advantage of the agency theory of causation is that of being a good substitute for the counterfactual theory. In this connection, Price argues that the agency theory not only conveys the same sort of counterfactual information conveyed by Lewis' theory, but it does it in a less controversial way, because it does not need any postulate on similarity among possible worlds and the like (see Price 1991).

In order to accept the agency theory one has to regard human beings as actors, not as pure observers. Therefore, this approach is not purely humean, but instead of appealing to non-Humean concepts which are modal or metaphysical in character, it embraces a pragmatist perspective which makes causality depend on the agent's knowledge and beliefs. However, as observed by Price, looking at causality as perspectival in character, or as "a manifestation of the fact that causal concepts originate in our experience as agents" (Price 1992a, p. 501), does not put it beyond the reach of science, nor does it rule out a realist account of causation, if one wants it. A realist, who wanted to interpret observed correlations as expressions of objective causal relations existing in the world, could still accept the agency view on the claim that agency is not to be taken as a constituent of the world, but rather as "what makes causation accessible and important for agents" (Price 1991, p. 173). Even taking for granted that there are objective causal relations in the world, the realist should be ready to admit that "as agents in the world, we are capable of exploiting these relations to further our ends" (p. 172).

Although the agency view does not conflict in principle with a realist explication of probabilistic causality, according to its proponents it "does serve to undermine a popular recent argument for causal realism" (Price 1991, p. 161), namely Cartwright's plea for non-Humean causality, which also appeals to some form of agency (effective strategies). In contrast, Price claims that "given the constraints of the agent's perspective, ordinary procedures of evidential reasoning can draw the distinctions they are said to be unable to draw. The distinction between effective and ineffective strategies needs evidential reasoning by agents, and nothing stronger" (*ibidem*). In other words, evidence, and in particular experimental relative frequencies, guide the agent's choice of those actions which, in the light of past experience, are effective strategies for achieving certain ends.

The perspective outlined by Price and Menzies provides a general philosophical framework for agency causality. In order to become a flesh and blood theory of causality, it has to be substantiated by more specific accounts. The literature on econometric model building offers a number of such, ranging from the classical approach developed by H. Simon and H. Wold—which takes causality in a manipulative sense and defines it with reference to (deterministic) models—to more recent accounts where models are specified in terms of statistical variables[2]. In both cases causality is associated with exogeneity and some sort of invariance. In nonexperimental disciplines like econometrics, characterized by a weaker theoretical status than physics, causality acquires a peculiarly pragmatical character. Models are typically context-dependent constructs, and their specification depends on theoretical assumptions underlying the choice of exogenous variables, as well as on the nature of the available data. In the case of statistical models, other elements come into play, like the expectations of economic agents. In addition, the specification of an econometric model usually depends on the purpose for which the model is to be used.

5   Towards an Integration

Let us examine the possibility of combining Salmon's mechanical approach with the agency theory of causation. A first advantage of such an integration would amount to the fact that causal talk would receive counterfactual import in a straightforward and noncontroversial way.

In spite of Price's claim that the agency theory of causation can be combined with a realistic interpretation of the correlations involved, the

---

[2] For a discussion of causality in econometrics see Galavotti and Gambetta (1990).

introduction of the agency theory within Salmon's approach brings with it a pragmatic flavour. This does not concern only the introduction of contextual considerations into causal analysis, recently accepted also by Salmon. The pragmatical character of agency causation involves an equally pragmatical interpretation of laws and theories, of the kind reaffirmed by Ramsey. In his commentary to my "Wesley Salmon on Explanation, Probability and Rationality" (Galavotti 1999), Salmon claims to be ready to accept an "epistemic approach to theories and laws" because he thinks that this "would not necessarily undermine the ontic status of causality" which he wants to retain. However, he also regards convergence of opinion, which is the criterion usually advocated by pragmatists to substantiate scientific truth, as too weak for a realist like him.

The "ontic status of causality" advocated by Salmon rests on the specification of causal mechanisms and the idea that homogeneous reference classes can be obtained. Even granted that in some fields this can be done, there are many areas where information on homogeneous reference classes cannot be reached and/or causal processes cannot be devised. A manipulative view of causality would be of some help in such cases, allowing recognition of causal properties even when the homogeneity requirement is not fulfilled. So, under this interpretation, Salmon's theory would find an easier application to such cases like those pointed out by Woodward in connection with the social sciences and mentioned above. If this were true, the combination of Salmon's model with a manipulative theory of causation would widen its range of application.

Another point stressed by Woodward is that in the social sciences the sought explanation of phenomena usually does not regard single case occurrences of events, but rather facts at the population level. This is testified by the extensive use of statistical techniques for regression analysis, which convey the kind of information that is typical of the S-R basis. According to Woodward, this is a further argument against the suitability of Salmon's mechanical approach to the social sciences.

These remarks bring us back to the problem of the two levels of causality. To a certain extent, the impact of this problem varies according to the context in which it arises. The mechanistic picture of causality advocated by Salmon suits disciplines like classical and statistical mechanics, which are characterized by a strong theoretical apparatus and by the use of general laws to describe the phenomena. In these fields, provided the available information allows identification of causal mechanisms, statistical causality will match token causality. In other disciplines, characterized by a weaker theoretical apparatus, the description of phenomena rests on models and the linkage between statistical and token causality, as well as that between explanation and prediction, becomes looser. This is the case with the social sciences and econometrics. So, within different scientific contexts the two kinds of causality stand in a different relation to each other.

As Price is aware, embracing the agency view is not sufficient to solve the problems connected to the existence of two levels of causality. However, Price holds the conviction that the agency approach can help to clarify some aspects of such problems[3]. We are led to an analogous conclusion by the literature on econometrics, where the shift between population and token causality acquires a peculiar significance, being linked to the problem of aggregation. In broad terms, this amounts to the problem of connecting the behaviour of single agents (at the level of micro-economics) to the behaviour of the so-called "representative agents" (at the level of macro-economics). With reference to causality, the problem is that of combining the causal structure of the microeconomic model, where exogenous variables are said to causally influence the endogenous variables, with the interdependence among the variables of the macroeconomic model, where the behaviour of each representative agent is simultaneously dependent on the behaviour of the others.

In his "Causality and Temporal Order in Macroeconomics or Why Even Economists Don't Know How to Get Causes from Probabilities" (1993) Kevin Hoover makes this problem the starting point of an attack against accounts of causality based on temporal order, in particular Cartwright's approach. He claims that what is needed instead is a manipulative approach to causation, and advocates Simon's view that "causal ordering is invariant to well-defined classes of interventions" (Hoover 1993, p. 706). He also conjectures (without developing the argument) that Simon's account "bears a family resemblance to Salmon's view that a causal process is one that will transmit marks" (*ibidem*). Actually, such an analogy does not seem to go a long way, beyond the fact that both accounts appeal to some sort of invariance. Between the two approaches there is a fundamental difference, lying in the deterministic nature of Simon's approach to causality, where models are not specified in terms of statistical variables.

The last two decades have seen an enormous outgrowth of econometric models specified in probabilistic terms. While retaining the manipulative notion of causality, they face peculiar problems with regard to the aggregation problem. The situation can be summarized as follows. In econometrics it is usually assumed that a particular economic agent, the policy maker, tries to influence the behaviour of other agents by manipulating the value of some exogenous variables, usually called instrument variables. Now, while it can be assumed that such variables are strictly causal with respect to single agents, in the sense that single agents cannot influence their value, at the

---

[3] The problem is discussed in Price (1991, 1992b).

aggregate level the behaviour of all agents can have feedback effects on the instruments of the policy maker, thereby altering the causal structure of the effect. In view of this, it is necessary to introduce a conditional concept of manipulability, according to which the efficacy of the intervention of the policy maker is subordinate to the body of information available to economic agents.

In general, the heterogeneity of individuals and the uncertainty due to their interaction on the market make it impossible to pass from individual to aggregate behaviour in any simple way (like summing up). Assuming that heterogeneity and uncertainty can be described by means of probability distributions, models will have to be specified in terms of relations between the expected values of such distributions, conditional on the body of information available to economic agents, where information varies not only over time, but also according to the chosen level of aggregation.

The conclusion attained by the literature on econometrics is that the relationship between causality referred to statistical variables and causality referred to individuals can be handled only at the price of introducing a conditional notion of manipulability and a dynamical specification of causal structure, dependent on the expectations and the information available to economic agents, as well as on the purpose for which models are used[4]. Clearly, the nature of the "individuals" forming the object of econometrics, namely economic agents, is at least in part responsible for the necessity of weakening even the notion of agency causality, in the sense suggested above. What happens in other fields dealing with natural phenomena is a matter for further investigation. The problem of connecting the two levels of causality remains open. Though embracing the manipulative view is not enough to solve it, its adoption might still help to clarify the matter, as seems to be the case with econometrics.

Salmon's causal mechanisms based on homogeneous reference classes and having an ontic status look quite distant from the pragmatical and context-dependent notion of causation linked with manipulation. However, the agency approach is flexible enough to accommodate stronger accounts, including Salmon's. In order to mend the fracture between the two approaches one would have to start from the weaker framework of the agency theory of causation, and keep the mechanical picture as an ideal. When scientific theories allow causation to be specified in terms of mechanisms, that ideal will be fulfilled, otherwise manipulability will suggest weaker accounts of causation. Still, the mechanistic picture can be retained as a heuristic device.

---

[4] For a discussion see Galavotti and Gambetta (1999).

References

Galavotti, M. C. (1999). Wesley Salmon on explanation, probability and rationality. In M.C. Galavotti and A. Pagnini (Eds.), *Experience, Reality and Scientific Explanation*. Dordrecht-Boston: Kluwer, pp. 39-54.

Galavotti, M. C. and Gambetta, G. (1990). Causality and exogeneity in econometric models. In R. Cooke and D. Costantini (Eds.), *Statistics in Science. The Foundations of Statistical Methods in Biology, Physics and Economics*. Dordrecht-Boston: Kluwer, pp. 27-40.

Galavotti, M. C. and Gambetta, G. (1999). Theory and observation in econometric models: a constructivist approach. In R. Rossini Favretti, G. Sandri and R. Scazzieri (Eds.), *Incommensurability and Translation*. Cheltenham-Northampton: Edward Elgar, pp. 339-349.

Good, I. J. (1961) and (1962). A causal calculus I and II. *British Journal for the Philosophy of Science*, 11, 305-318; 12, 43-51; Errata and corrigenda, 13, 88.

Hitchcock, C. R. (1995). Discussion: Salmon on explanatory relevance. *Philosophy of Science*, 62, 304-320.

Hoover, K. D. (1993). Causality and temporal order in macroeconomics or why even economists don't know how to get causes from probabilities. *British Journal for the Philosophy of Science*, 44, 693-710.

Menzies, P. and Price, H. (1993). Causation as a secondary quality. *British Journal for the Philosophy of Science*, 44, 187-203.

Price, H. (1991). Agency and probabilistic causality. *British Journal for the Philosophy of Science*, 42, 157-176.

Price, H. (1992a). Agency and causal asymmetry. *Mind*, 101, 501-520.

Price, H. (1992b). The direction of causation: Ramsey's ultimate contingency. In D. Hull, M. Forbes and K. Okruhlik (Eds.) PSA 1992, Vol. 2. Philosophy of Science Association, pp. 253-267.

Price, H. (1996). *Time's Arrow and Archimedes' Point*. New York-Oxford: Oxford University Press.

Reichenbach, H. (1956). *The Direction of Time*. Berkeley and Los Angeles: University of California Press.

Salmon W. C. (1971). Statistical explanation. In: W. C. Salmon, R. C. Jeffrey and J. G. Greeno, *Statistical Explanation and Statistical Relevance*. Pittsburgh: University of Pittsburgh Press.

Salmon, W. C. (1984). *Scientific Explanation and the Causal Structure of the World*. Princeton: Princeton University Press.

Salmon, W. C. (1994). Causality without counterfactuals. *Philosophy of Science*, 61, 297-312. Also in Salmon (1998), pp. 248-260.

Salmon, W. C. (1997). A reply to two critiques. *Philosophy of Science*, 64, 461-477.

Salmon, W. C. (1998). *Causality and Explanation*. New York-Oxford: Oxford University Press.

Suppes, P. (1970). *A Probabilistic Theory of Causality.* Amsterdam: North-Holland.

Suppes, P. (1984). Conflicting intuitions about causality. In French et al. (Eds.), *Midwest Studies in Philosophy Volume IX. Causation and Causal Theories,* Minneapolis: University of Minnesota Press, pp. 151-168.

Woodward, J. (1989). The causal mechanical model of explanation. In P. Kitcher and W. C. Salmon (Eds.), *Scientific Explanation.* Minneapolis: University of Minnesota Press, pp. 357-383.

2

Metaphysics over Methodology—or, Why Infidelity Provides No Grounds to Divorce Causes from Probabilities

DAVID PAPINEAU

King's College London

1 Introduction

A reduction of causation to probabilities would be a great achievement, if it were possible. In this paper I want to defend this reductionist ambition against some recent criticisms from Gürol Irzik (1996) and Dan Hausman (1998). In particular, I want to show that the reductionist programme can be absolved of a vice which is widely thought to disable it—the vice of infidelity.

This paper also carries a general moral. It is dangerous to muddle up metaphysics with methodology. If you are interested only in the methodological question of how to *find out* about causes, you will be unmoved by my defence of reductionism, since it hinges on metaphysical matters that are of no methodological consequence. Indeed, if you are interested only in methodological matters, you may as well stop reading here, since my reductionism will offer no methodological improvement over the non-reductionist alternatives.

On the other hand, if you are interested in the underlying structure of the universe—and in particular in how there *can* be causal direction in a world whose fundamental laws are symmetrical in time—then I may have

Stochastic Causality.
Maria Carla Galavotti, Patrick Suppes and Domenico Costantini (eds.).

something for you. I admit my favoured theory offers nothing new to market researchers who want to find out whether some form of advertising causes improved sales. But I can live with that, if my theory explains the arrow of time.

For any readers new to this area, I should explain that the kind of reductive theory at issue here only has *partial* reductive ambitions, in that it takes *probabilistic* laws as given, and then tries to explain *causal* laws on that basis. The hope is to weave the undirected threads of probabilistic law into the directed relation of causation. Perhaps it would be more helpful to speak of a reduction of causal direction, rather than of causation itself. If we think of causal laws as being built from two components—first a symmetrical lawlike connection linking effect and cause, and second a causal 'arrow' *from* cause *to* effect—then the reductive programme at issue here aims only to reduce this second directional component. In particular, it aims to reduce it to facts involving the first kind of component (more specifically, to undirected probabilistic laws between the cause, effect *and* other events). However, it does not aim to explain these probabilistic laws themselves. (It says nothing, for example, about the difference between laws and accidental frequencies).

Most recent discussion of the reductionist programme has focused on methodological rather than metaphysical issues (including Papineau 1993a). This is understandable, to the extent that it is the possibility of real-life inferences from correlations to causes which motivates most technical work in this area. My strategy in this paper will be to place this methodological work in a larger metaphysical context. This metaphysical context won't make any difference to the methodology, but, as I said, methodological insignificance seems a small price to pay, if we can explain why causation has a direction.

2 Probabilistic Causation and Survey Research

A good way to introduce the metaphysical issues will be to say something about 'probabilistic causation' generally. Over the past three or four decades it has become commonplace to view causation as *probabilistic*. Nowadays, the paradigm of a causal connection is not C determining E, but C *increasing the probability* of E—$P(E/C) > P(E)$. (Suppes 1970.)

This shift in attitudes to causation is often associated with the quantum mechanical revolution. Given that quantum mechanics has shown the world to be fundamentally chancy, so the thought goes, we must reject the old idea of deterministic causation, and recognize that causes only fix quantum mechanical chances for their effects.

However, there is another way of understanding probabilistic causation, which owes nothing to quantum metaphysics. This is to hold that probabilistic cause-effect relationships arise because our knowledge of causes is incomplete. Suppose that C does not itself determine E, but that C in conjunction with X does. Then P(E/C) can be less than one, not because E is not determined, but simply because X does not occur whenever C does.

A useful label for this possibility is 'pseudo-indeterminism' (Spirtes, Glymour and Scheines 1993). If we focus only on C, and ignore X, it will seem as if E is undetermined. But from a perspective which includes X as well, this indeterminism turns out to be illusory.

This pseudo-indeterministic perspective in fact fits much better with intuitive thinking about 'probabilistic causation' than quantum metaphysics. The real reason contemporary intuition associates causes with probabilities is nothing to do with quantum mechanics. (Indeed, when we do look at real microscopic quantum connections, causal ideas tend to break down, in ways I shall touch on later.) Rather, all our intuitively familiar connections between probabilistic and causal ideas have their source in *survey research,* or less formal versions of such research—and to make makes sense of these research techniques, we need something along the lines of pseudo-indeterminism, not quantum mechanics.

Let me explain. By 'survey research' I mean the enterprise of using statistical correlations between macroscopic event types to help establish causal conclusions. To take a simple example, suppose that good exam results (A) are correlated with private schools (B). Then this is a prima indication that schools exert a causal influence on exam results. But now suppose that in fact private schools and good exam results are correlated only because both are effects of parental income (C). If that is so, then we would expect the school-exam correlation to disappear when we 'control' for parental income: among children of rich parents, those from state schools will do just as well in the exams as those from private schools; and similarly among the children of poor parents.

In this kind of probabilistic case, C is said to 'screen off' A from B. Once we know about C (parental income), then knowledge of B (school type) no longer helps to predict A (exam results). Formally, we find that the initial correlation—$P(A/B) > P(A)$—disappears when we condition on the presence and absence of C: $P(A/B\&C) = P(A/C)$ and $P(A/B\&\text{-}C) = P(A/\text{-}C)$.

To continue with this example for a moment, focus now on the correlation between parental income and exam results itself. Suppose that survey research fails to uncover anything which screens off this correlation, as parental income itself screened off the initial correlation between schools and exam results. Then we might on this basis conclude that parental income *is* a genuine cause of exam results.

Inferences like these are commonplace, not just in educational sociology, but also in econometrics, market research, epidemiology, and the many other subjects which need to tease causal facts out of the frequencies with which different things are found together.[1] Now, it is a large issue, central to this paper, whether any causal conclusions ever follow from such statistical correlations alone, or whether, as most commentators think, statistical correlations can only deliver new causal facts if initially primed with some old ones ('no causes in, no causes out'). But we can put this issue to one side for the moment. Whether or not survey research requires some initial causal input before it can deliver further causal output, the important point for present purposes is that, when survey research does deliver such further conclusions, these conclusions never represent purely chance connections between cause and effect.

Suppose, as above, that survey research leads to the conclusion that parental income is a genuine cause of exam results. Now, the soundness of this inference clearly doesn't require that *nothing else* make a difference to exam results, apart from parental income. For parental income on its own clearly won't fix a pure chance for exam results. Other factors, such as the child's composure in the exam, or whether it slept well the night before, will clearly also make a difference. All that will have been established is that parental income is *one* of the factors that matters to exam results, not that it is the *only* one. As it is sometimes put, parental income will constitute an '*in*homogeneous reference class' for exam results, in the sense that different children with the same parental income will still have different chances of given exam results, depending on the presence or absence of other factors. ($P(E/C \textit{ and } X) \neq P(E/C \textit{ and not-}X)$.)

This point is often obscured by worries about '*spurious*' correlations. If we want to infer, from some initial correlation between C and E, that C *causes* E, we do at least need to ensure that the C-E correlation doesn't disappear when we 'control' for possible common causes. The point of survey research is precisely to check, for example, whether or not the parental income-exam correlation can be explained away as due to the spurious action of some common cause. Thus in practice we need to check through all pos-

---

[1] Such research is usefully thought of as proceeding in two stages. First, we need to get from finite sample data to lawlike probability distributions; second, we need to get from these probability distributions to causal structure. The first stage hinges on standard techniques of statistical inference. In this paper I shall say nothing about the logic of such techniques, important as this subject is, and simply assume knowledge of lawlike probability distributions. My focus here is exclusively on the second stage, which takes us from lawlike probabilities to causal conclusions. This division of labour is in line with the limited ambitions of my reductive agenda: as I explained at the beginning, the aim is only to reduce causal direction to lawlike connections, not to reduce the latter in turn.

sible common causes of C and E, and make sure C still makes a difference to E *after* these are held fixed. This might make you think that survey research needs to deal in pure homogeneous chances after all. For haven't I just admitted that we are only in a position to say C causes E when we know the probability of E given C and X1 . . . Xn, when these Xs are all the other things which make a probabilistic difference to E?

No. I said we need to check for all possible *common* causes of C *and* E. I didn't say we need to check through *all* other causes of E tout court. This difference is central to the logic of survey research. Before we can infer a cause from a correlation, we do indeed need to see what difference any common causes make to the probability of E. But we don't need to know about every influence on E. This is because most such influences will be incapable of inducing a spurious correlation between C and E. In particular, this will be true whenever these other influences are themselves probabilistically independent of the putative cause C. If some other cause X (good night's sleep) is probabilistically independent of C (parental income), then it *can't* generate any spurious C-E correlation: X will make E more likely, but this won't induce any co-variation between C and E, given that C itself doesn't vary with X. So survey research can happily ignore any further causes which are probabilistically independent of the cause C under study. The worrisome cases are only those where the further cause X is itself correlated with C, since this will make C vary with E, even though it doesn't cause E, because it varies with X which does.

The moral is that you don't need to gather statistics for every possible causal influence on E whenever you want to use survey data to help decide whether C causes E. You can perfectly well ignore all those further influences on E (all those 'error terms') that are probabilistically independent of C. And of course this point is essential to practical research into causes. In practice we are never able to identify, let alone gather statistics on, all the multitude of different factors that affect the Es we are interested in. But this doesn't stop us sometimes finding out that some C we can identify is one of the causes of E. For we can be confident of this much whenever we find a positive correlation between C and E that remains even after we hold fixed those other causes of E that C *is* probabilistically associated with.[2]

---

[2] I do not of course want to suggest that it will be trivial, or easy, or even something we can ever be fully certain about, to identify all other causes X of E that some putative cause C is correlated with. But it is certainly far easier than identifying all causes of E tout court. Moreover, in practice, we can often use background knowledge to attain a fair degree of confidence. Note also how *randomized trails* exploit the difference between uncorrelated Xs, which we can ignore, and correlated ones, which must be taken into explicit account. The effect of randomizing a 'treatment' C is precisely to push all other Xs into the former category, by forcibly decorrelating them from C. (Cf. Papineau 1989, 1993b.)

The important point in all this is that familiar cases of 'probabilistic causation' are nothing to do with pure quantum mechanical chances. In typical cases where C 'probabilistically causes' E, the known probability of E given C will not correspond to any chance, since C will not constitute a homogeneous reference class for E.

Note what this means for the significance of conditional probabilities. When survey research shows us that P(E/C) is greater than P(E/-C), and that this correlation is non-spurious in the sense that it remains when we condition on further variables, this does not mean that C alone fixes that chance for E. Nor does it even mean that C, in conjunction with whichever other Xs are present in given circumstances, always increases the chance of E by the difference P(E/C) - P(E/-C). For it may be that C *interacts* with some of these other Xs, making different differences to the chance of E in combination with different Xs, or perhaps even *decreasing* the chance of E in combination with some special Xs. All the non-spurious P(E/C) - P(E/-C) implies is that C rather than not-C makes that much difference to the chance of E *on weighted average over combinations of presence and absence of those other Xs* (with weights corresponding to the probability of those combinations).[3]

3 Pseudo-Indeterminism and Common Causes

Now, these points do not yet constitute an argument for 'pseudo-indeterminism'. It is one thing to argue that survey research always involves unconsidered 'error terms' which make further differences to the chances of effects. It is another to hold that, when these 'error terms' are taken into account, the chances of effects are then always zero or one. This would not only require further error terms which make *some* differences to the chances of effects; in addition, these further differences must leave all chances as zero or one.

---

[3] In the view of Ellery Eells, 'Average effect is a sorry excuse for a causal concept' (1991, p. 113). Let me make two points. First, this causal concept doesn't seem so sorry in connection with rational action, since in that context it is exactly what we need. We don't normally know all the details of our situation, and in such cases rationality dictates precisely that we should perform an action C in pursuit of E just to the extent that C non-spuriously increases the probability of E on weighted average over all the situations we might be in. (Beebee and Papineau 1997; Papineau forthcoming.) And, second, the average effect concept should not in any case be thought of as a primitive concept of causation, since it is nothing but an average over the different ways C might operate as a *single-case* cause of E in particular cases. I take C to be such a single-case cause if the chance of E in the actual circumstances is higher than it would have been if C had been absent. (This is not to deny that these single-case counterfactuals might be reducible to generic causal laws and initial conditions; but the generic laws needed here will need to be more fine-grained than statements of C's 'average effect' on E.)

Still, I think there is some reason to hold that just such deterministic structures lie behind the causal relationships we are familiar with. This relates to a feature of common causes discussed in the last section. Recall how common causes 'screen off' correlations between their joint effects: the joint effects will display an initial unconditional correlation, but conditioning on the presence and absence of the common cause renders them uncorrelated.[4]

Now, one interesting feature of this screening-off property is that it illustrates how the probabilistic relationships manifested by causes and effects are arranged asymmetrically in time. Note how the probabilistic screening-off relation between common causes and joint effects is absent from the 'causally-reversed' set-up where a common effect has two joint causes. We don't generally find, with a common effect (heart failure, say), that different causes (smoking and over-eating, say) are correlated; moreover, when they are, we don't generally find that the correlation disappears when we control for the presence and absence of the common effect.[5]

When I first started working on the direction of causation, I thought that this asymmetry might provide the key (Papineau 1985b). Think of the problem as that of fixing the causal arrows between a bunch of variously correlated variables. Just knowing which variables are pairwise correlated clearly won't suffice, since pairwise correlation is symmetrical—if A is correlated with B, then B is correlated with A. But if common causes differ from common effects in respect of screening-off, then perhaps we can do better, and can mark C down as a common cause of joint effects A and B, rather than an effect of A or B, whenever we find a C that screens off a prior correlation between an A and B.

In fact, though, this is too quick. For screening off does not itself ensure that C is a common cause. The screening-off probabilities are also displayed

---

[4] This is in fact a oversimplification. In more complex causal structures, for example where A and B have two common causes, neither common cause will screen off the A-B correlation by itself. (Hausman 1998, p. 209.) A similar point applies to the screening-off property of intermediate causes. I shall skate over this complication in what follows, as it will not matter to the overall argument.

[5] It should be noted, though, that in particular cases controlling for the presence (but not absence) of a common effect will have precisely the effect of screening off a prior correlation between joint causes. This is because controlling for a common effect serves to induce a negative correlation between its joint causes (Hausman, 1998, p. 83). If the numbers are right, this effect can thus cancel out a pre-existing correlation between joint causes. (Cf. Irzik, 1996, sect 5.) Irzik raises this point because it disproves a claim about the connection between causes and probabilities that I made in earlier work (1993a). However, it does not, so far as I can see, affect the version of reductionism outlined in section 4 below. For some further explanation of why controlling for common effects induces negative correlations between joint causes, see footnote 8 below.

when C is causally intermediate between A and B (thus A→B→C, or B→C→A). So a probabilistic 'fork' (C screens off A from B), doesn't guarantee that C is a common cause. C could also be causally intermediate between A and B. Still, even this gives us something to work with. When we find a probabilistic fork, we can at least rule out C's being a common effect (A→C←B), and be confident that one of the other three possibilities applies. And then, perhaps, by repeatedly applying this inference to different triples from the overall bunch of variables, we might be able to determine a unique ordering among them all.[6]

In the end, however, the screening off asymmetry turns out to be less central to the reductionist programme that I originally supposed. In the next section I shall borrow from Dan Hausman's work (1998) to lay out the basic requirements for a reduction of causation to probabilities. From the perspective there developed, the important requirement is not so much that common and intermediate causes should screen off unconditional correlations, but rather that there should be probabilistically independent causes for any given effect. This requirement by itself is enough to tell us, for any correlated A and B, whether A causes B, B causes A, or whether they are effects of a common cause. Relative to this basic independence requirement, screening off only plays the relatively minor role of distinguishing direct from indirect causes (and indeed the screening off property of common causes, as opposed to that of intermediate causes, seems to play no important role at all).

Still, there remains an important connection between Hausman's basic independence requirement and the screening-off property. If we conjoin the independence requirement with the hypothesis of pseudo-indeterminism, then we can *explain* the screening-off property, when otherwise it must be taken as primitive. This returns me to the main theme of this section. I want to argue for pseudo-indeterminism (that is, the existence of underlying deterministic structures), on the grounds that we need pseudo-indeterminism to explain the phenomenon of screening off.

It is worth being clear what I am aiming to explain here. My idea isn't to explain why certain *causal specificed structures* display probabilistic screening off. From my perspective, this connection holds as a matter of metaphysical necessity. Though the suggestion has yet to be made good, I am assuming that probabilistic screening-off is part of what constitutes certain events as being causally interposed between others. Rather, my target here is to explain why there should be any events that are related by screening off in the first place—that is, why we should find any triples such

---

[6] This general idea goes back to Reichenbach (1956).

that A and B are unconditionally correlated, yet the correlation disappears when we control for C.

At bottom, the idea is simple. Let me explain it in connection with common causes. I shall return to intermediate causes at the end of the section. Suppose that some common cause C has two effects A and B. Suppose further that there are 'error terms' X and Y, such that $(CX \text{ or } Y) \leftrightarrow A$. That is, C is a deterministic 'INUS' condition for A: whenever A occurs, either C&X determines it, or some other Y does. Now suppose similarly that C is a deterministic INUS conditions for B: there are further 'error terms' S and T such that $(CS \text{ or } T) \leftrightarrow B$. Then, provided the relevant error terms are probabilistically independent of each other, this guarantees that there will be an unconditional correlation between A and B, which will be screened off by C.

Intuitively, if you've got A, then this adds to the probability that C was there to produce it, and so, since C also makes B more likely, this will generate a resulting correlation between B and A. (More carefully, this inference will be valid precisely insofar as the factors which ensure B in the presence of C—namely, S-or-T—are not negatively correlated with A.)

Then, to get the screening-off, note that, among cases where C is present, X-or-Y and S-or-T will necessary and sufficient for A and B respectively—and among cases where C is absent, Y and T will be similarly necessary and sufficient for A and B respectively. So, as long as these error terms are appropriately probabilistically independent, A and B will also be probabilistically independent given C and not-C. (Cf. Papineau, 1985b.)[7]

By way of confirmation of this story, note how it will not work 'in reverse' to predict screened-off correlations among deterministic joint causes of common effects. For in these 'backwards' cases it is clear that the background conditions will not satisfy the requisite independence requirements. To return to our earlier example, it is not impossible that there should be some X and Y such that a heart attack (H) plus X, or Y, is coextensive with earlier smoking. (Think of X as signs that the heart attack comes from smoking, and of Y as non-heart-attack traces of smoking.) Similarly, H&S or T might be necessary and sufficient for earlier overeating. But we won't on this account expect to find that overeating is correlated with smoking, precisely, because given a heart attack, overeating can be expected to be negatively correlated with X or Y, and in particular with X, since in cases of

---

[7] Note how the 'error terms' will *not* be probabilistically independent, and so not ensure screening off, in the case where there is another common cause for A and B apart from C (cf. footnote 4). For in these cases some of the non-C-involving causes of A and B will themselves be correlated, due to this further common cause.

overeating it is less likely that there will signs that the heart attack has been preceded by smoking.[8]

Now, one natural reaction to this suggested explanation of common causal screening-off would be 'Sure, that's what we should believe if determinism were true. But determinism isn't true.'

I agree that determinism is not true. Quantum mechanics shows that a full specification of current circumstances will often fail to determine what happens next. Still, it is an important that my suggested explanation of common causal screening-off does not presuppose universal determinism, according to which every occurrence is fixed by prior circumstances. Indeed, the requirements of my explanation are so weak that it is somewhat misleading to characterise them as 'deterministic' at all. All the explanation requires is that there be further factors which, together with C, fix the occurrence of A and B respectively, and that these further factors should be probabilistically independent. It does not require that these further factors, or C for that matter, should themselves be determined by anything. Nor indeed does it require that these further factors be contemporaneous with C: They could equally well be circumstances which emerge chancily in the interval between C and its effects.

Still, this is unlikely to satisfy my objector. Maybe my story doesn't imply universal determinism. But I can still be pressed on why we should suppose even the limited kind of deterministic structure at issue. I earlier made the point that the familiar logic of survey research commits us to the existence of unknown 'error terms'. But, as I admitted at the start of this section, this in itself falls short of the further contention that these 'error terms' will restore determinism by fixing chances of zero or one for all results. And as yet I have offered no real evidence for this further contention. Why then suppose that all screening-off common causes must be associated with 'error terms' in conjunction with which they fix their effects? After all, if quantum mechanics shows that some earlier events fix chances other than zero or one for later events, then surely I should allow that some screening-off common causes will similarly relate in a purely chancy way to their joint effects? My earlier points about survey research seem to leave it quite open that there should be purely 'quantum mechanical common causes' Cs which fix pure chances for correlated joint effects A and B.

---

[8] This now helps us to understand why controlling for the presence of common effects should induce negative correlations among joint causes (cf. footnote 4). In the example, overeating becomes a negative indicator of smoking, once the heart attack is taken as given—for once we know that the heart attack came from overeating, we have less reason than before to suspect the victim of smoking.

I agree that this possibility is open. At the same time, it is one of the most striking features of quantum mechanics that, when we look for cases of prior quantum mechanical states that fix pure chances for correlated events A and B, we find that such prior quantum states characteristically fail to screen off these correlations, and so fail to display the probabilistic structure constitutive of common causes. (What is more, it can be shown that many such cases involves structures of correlations which cannot be screened-off by *any* local prior states, even if they are different from those prior states recognized by orthodox quantum mechanics.)

I am here thinking of the well-known 'EPR' correlations. This is not the place to add to the large literature on this subject. My point is simple enough. Once quantum mechanics persuades us of indeterminism, it seems natural enough to suppose that some prior Cs will simultaneously fix pure chances for two events A and B, in such a way that A and B are correlated and their correlation will screened of by C. But, in fact, when we look into quantum mechanics, we don't normally find pure-chance-fixing prior states with this screening-off characteristic. Moreover, this can't be put down to the temporary failings of current quantum theory, since the relevant correlations often have a collective correlational structure which cannot possibly be screened off by any local prior states.

There isn't of course anything conceptually impossible about some C on its own fixing pure chances for A and B, such that A and B are correlated and C screens off the correlation. But what the EPR cases bring home is that there is nothing conceptually *inevitable* about this either. That is, it is perfectly possible to have some prior state C which fixes pure chances for A and B, such that A and B are correlated, but where C *doesn't* screen off this correlation.

Compare this with the deterministic case where C&X or Y determine A, and C&S or T determine B. Given the independence requirements on the 'error terms', this deterministic set-up *guarantees* that C will screen off the correlation between A and B. In the purely chancy case, by contrast, there is nothing which forces C to screen-off the A-B correlation.

If we look at things from this perspective, it is possible to be less surprised than many people are by the fact that quantum mechanics gives rise to strange correlations that can't possibly be screened off by common causes. Think of it like this. If a prior factor C fixes two later results via some underlying deterministic structure, then, given independence requirements, there is no possibility of the correlation between the later results not being screened off by C. But if we don't have this underlying deterministic structure, as in purely chancy quantum mechanical situations, then there is probabilistic room, so to speak, for A and B to become correlated in a way that isn't screened off by prior circumstances. And it turns out that, as soon as nature has this room, it uses it to produce unscreenoffable correlations. What's so surprising about that?

So far I have only defended this pseudo-indeterministic explanation of screening off in connection with common causes. But the same story could be told for intermediate causes (A→C→B). Provided there are 'error terms' together with which such Cs determine their Bs, and provided these are probabilistically independent of the corresponding backwards 'error terms' which fix whether or not the Cs come from the As, then we will have an explanation for the screening-off property of intermediate causes quite analogous to that given for common causes, and with all the same virtues.

Perhaps it is worth emphasising once more that these arguments only need a very minimal 'determinism' in order to render the apparent 'indeterminism' as 'pseudo'. What we need are laws of the form C&X v Y ↔ A, where X and Y are independent of the similar factors together with which C fixes other effects and causes. As I said earlier, this doesn't require that X or Y or C are themselves determined. Nor does it require that X or Y be contemporaneous with C. It would be enough, for instance, if there were a plethora of chancy microscopic occurrences temporally between C and A, some of which 'helped' C to fix A, some of which 'hindered' this, and some of which had the power together to fix A even without C. Provided these microscopic events plus C collectively fix whether or not A occurs, *and* provided they are probabilistically independent of similar factors relevant to the other effects and causes of C, then we will get the probabilistic screening-off structure.

4   Bayesian Nets

So far I have argued for a certain metaphysical picture. Behind the variables involved in normal examples of probabilistic causation are further error terms which also make a difference to the chances of effects. Survey research can ignore these error terms, provided they are probabilistically independent of the causes under explicit study. Moreover, if we assume that these independent error terms suffice to determine effects and causes, then we can explain the screening-off property displayed by common and intermediate causes. Since quantum mechanics suggests that this screening-off property would not be displayed without such deterministic structures, we thus have reason to accept that such structures underlie the apparent indeterminism of familiar probabilistic causes.

In the rest of this paper I want to use this metaphysical picture to respond to some standard objections to the reductionist programme. A first task, however, is to explain how this reduction might work. So far I have alluded to the possibility of reducing causes to probabilities, without actually explaining this.

We think of the problem like this. Suppose we have some set of variables V, together with their joint probability distribution. We know about the correlations between any pair of variables in V conditional on any others. The question is whether this information serves to fix a causal order among those variables. Do the correlations determine which variables should be linked by direct causal arrows?

There has been a great deal of work on this problem over the past decade under the heading of 'Bayesian net'[9] theory, particularly from research groups led by Judea Pearl and Clark Glymour respectively (Sprites, Glymour and Scheines 1993; Pearl and Verma 1994). As I pointed out earlier, it is clear that not any correlations among any set of variables will serve to fix a causal direction. (If C screens off an initial correlation between A and B, then this in itself won't decide between: (i) A causes B through C, (ii) B causes A through C, or (iii) A and B are joint effects of C.) Still, the 'Bayesian net' research has shown that such causal ambiguities can always be resolved by more more complex sets of correlations. perhaps involving further variables. (Cf. Hausman, 1998, pp. 211-4.)

Such inferences from correlations to causes hinge on *assumptions* about the relationships between correlations and causes. In the literature, these come under various titles, the most familiar of which is the 'Causal Markov condition' of Spirtes, Glymour and Scheines. Here I would like appeal to an alternative version of such assumptions developed by Dan Hausman in his *Causal Asymmetries*.

As it happens, Hausman himself doesn't aim to reduce causes to correlations, but to a primitive notion of 'nomic connection', for reasons to which I shall turn in later sections. However, those who do seek a reduction to correlations, like myself, will do well to mimic Hausman's elegant reductive strategy, simply substituting 'probabilistic correlation' for 'nomic connection' where necessary. The resulting theory does a great deal to clarify the issues. In particular, if we follow Hausman, we can proceed in stages, first specifying assumptions which allow us to move from correlations to decisions about which variables causes which, but which don't decide whether such causal links are direct or indirect. Further assumptions then allow us to discriminate direct from indirect causes.

---

[9] I am unclear about the rationale for this terminology. It is not obviously appropriate, given that the subject has nothing to do with personal probabilities as such, nor with updating them by conditionalization. True, such updating of personal probabilities provides one good theory of statistical inference. But, as explained in footnote 1, questions of statistical inference are best kept separate from questions about inferring causes from lawlike probabilities.

When we divide things up in Hausman's way, it turns out that the basic source of causal direction is not the screening-off asymmetry on which I originally focused, but simply the requirement that effects should have probabilistically independent causes. This latter requirement is all we need at first pass, when we are only aiming to decide what causes what. Probabilistic screening off only matters at the second step, when we need to ascertain whether A causes B directly or through some intermediary.

Hausman shows (modulo my substitution of 'correlations' for 'nomic connections') that if we assume

(I) A and B are correlated *if and only if* either A causes B, or B causes A, or they have a common cause

(II) if A causes B, or A and B have a common cause, then B is correlated with something that is probabilistically independent of A

then it follows that

(III) A causes B *if and only if* A and B are correlated, and everything correlated with A is correlated with B, and something correlated with B isn't correlated with A.

The two premises involved here will be examined in detail in the following two sections. But I trust they strike readers as having at least some intuitive appeal.

The first premise (I) is simply the idea that any correlation must have a causal explanation, combined, in the other direction, with the thought that causal connections will show up in correlations.

The second premise (II) may be less familiar, but also has some intuitive plausibility. It requires only that whenever some A causes some B, or they have a common cause, there will always be some further influence on B (think 'error terms') which is probabilistically independent of A.

Together these two premises mean that we can tell effects from causes simply by noting that the effects have independent sources of variation. Given an A and B that are correlated, B can't cause A if it co-varies with something which is probabilistically independent of A. To see why this works, note that all the factors correlated with a given cause (any of *its* causes or effects or symptoms) will be correlated with any further effects it has. So B can't possibly cause A, if it covaries with something which isn't correlated with A. Similar reasoning shows they can't have a common cause if everything correlated with A is correlated with B.

Once Hausman has fixed arrows of direct-or-indirect causation in this way, then it is fairly straightforward to decide which of these arrows are direct and which indirect. Here we need only assume that, if A is only an indirect cause of B, then its initial correlation with B will be screened off by the conjunction of the *other* causes of B. Conversely, if the A-B correlation is not so screened off by B's other causes, then A must be a direct cause of B.[10]

Hausman's story thus offers an explicit reduction of the relevant causal relationships. Conclusion (III) above specifies a necessary and sufficient condition for direct-or-indirect causation solely in terms of correlations. And, given this, we can then explicitly specify that A *directly* causes B if and only if it directly-or-indirectly causes it and the A-B correlation isn't screened off by any of B's other direct-or-indirect causes. (Note that the reference to 'direct-or-indirect causes' in this latter specification can be eliminated via (III)).

Now, this reduction is only as good as the assumptions from which it follows. If these assumptions are doubtful, then so is the reduction. In the next section I shall look at assumption (II), which specifies that effects B always have sources of variation that are independent of given causes A. In the following section I shall examine (I), which requires causal connections always to be manifested in correlations, and vice versa.

5 Including the Right Variables

According to (II), whenever A causes B, there is some further X which varies with B but not A. Is this generally true?

If we approach this issue with methodological spectacles on, (II) can seem highly problematic. Suppose that we are conducting some survey, and have chosen to focus on some specific set of variables V. Then, for Hausman's principles to deliver the right answers, B must be represented as being correlated with some A-independent source of variation, for all A and B in V where B is not a cause of A. But there is no reason to suppose that *this* requirement will automatically be satisfied. Even if reality contains such independent sources of variation, our chosen set V may simply fail to include them. The upshot may be that Hausman's techniques fail to determine a causal order among the variables in V, or, even worse, that they determine a causal order which is different from the one that obtains in reality.

---

[10] Note how this only requires screening-off by intermediate causes, not by common causes. As far as I can see, a Hausman-style reduction does not need screening-off by common causes. The assumption that common causes screen off thus puts extra constraints on cause-probability relations, beyond the minimal constraints needed for probabilities to fix causal order. In particular, it implies that a common cause that does not screen off is not the only common cause. This implication can be methodologically significant in pointing researchers to unobserved causes.

This might seem to undermine the reductive promise of Hausman's strategy. For now it seems that his relationships of probabilistic dependency and independency are only guaranteed to fix the right causal order for certain selections of variables V, namely, those which include independent sources of variation for any *effects* that they represent.[11] However, if we have to use unreduced causal notions in this way to specify the conditions under which probabilities will determine casual structure, then the overall Hausmanian package will clearly fail to show that probabilities alone determine causal structure. (Cf. Irzik 1996, sect 8.)

However, to argue in this way is to confuse methodology with metaphysics. It is true that in practice survey researchers will always work with limited sets of variables V, and that 'Bayesian net' methods will therefore lead them astray if these sets do not satisfy the requirement that all effects are represented as having independent sources of variation. But this is a methodological matter. From a metaphysical point of view, all that matters is that *reality* should satisfy this requirement. It doesn't matter if certain subsets of variables present a misleading picture of causal structure, as long as reality itself does not. The important metaphysical question is whether God can read the causal facts off from the correlational ones, not whether limited human researchers can do this.[12]

Critics like Irzik are aware that reductionists will seek to defend their programme by switching the focus from methodology to metaphysics. But they see no reason to suppose that the metaphysical realm will fill the gaps exposed by methodological incompleteness. If the Vs used by real researchers are not guaranteed do determine the correct causal order, then what rea-

---

11 How does this relate to the requirement, stressed in section 3 above, that a survey should include any common causes of the variables under study? Well, showing that a causally prior C screens off a correlation between some A and B is one way of showing that A and B must have mutually independent sources of variation, and so can't be related as cause and effect. But it is an interesting corollary of 'Bayesian net' research that we can know this directly of some correlated A and B, even when we haven't identified any common cause, as a result of explicitly identifying independent sources of variation for both A and B. If A varies with something that doesn't vary with B, then A can't cause B; and if B varies with something that doesn't vary with A, then B can't cause A either.

12 In practice, survey researchers standardly add unobservable independent 'error terms' to their original sets V, and thereby specify probabilistic structures which do fix causal structure. But in general they only know where to put these error terms as a result of already knowing that certain variables do not cause others. This, I take it, is the source of the widespread consensus that 'no causes in, no causes out'. Still, this limitation would not apply to a being who could observe the independent 'error terms' directly. (Actually, even survey researchers aren't as badly off as you might suppose, since they often use temporal order to infer that certain variables can't cause others. Metaphysicians will want an account of causal order that does not presume temporal order (Papineau 1985b) but there is no reason why practical researchers should hobble themselves in this way.)

son is there to suppose that simply switching to larger Vs will remedy this failing? Might not even God be stuck with an inadequate V? (Irzik 1996, sect 7.)

I take the arguments in the first half of this paper to answer this challenge. They show that there are many probabilistically independent 'error terms' behind the macroscopic variables studied in real survey research. Moreover, it seems that these error terms need to be sufficiently pervasive to restore a kind of determinism, if the screening off phenomena we observe are to be explicable.

Given this, there seems plenty reason to suppose that God's V, as opposed to those used by limited humans, will satisfy Hausman's requirement (II). On my picture, reality is far more complex than it appears to survey researchers. Alongside the macroscopic variables that are salient to human beings, many unobserved influences also enter into the laws relating different events. Any given variable will thus be correlated with a multitude of others. From a God's-eye point of view, there will be no shortage of variables available to display the independence relationships required to fix causal direction.[13]

6   Causes and Correlations

Let me now turn to Hausman's other assumption:

(I)   A and B are correlated *if and only if* either A causes B, or B causes A, or they have a common cause.

This can be queried in both directions. First, aren't there correlations which are of no causal significance? Second, aren't there causal connections which fail to manifest themselves as correlations?

It is these worries that makes Hausman himself stand off from a full-blooded reduction of causal direction to correlations, and settle instead for a reduction to a primitive notion of 'nomic connection'. In Hausman's view, while nomic connections and correlations normally go together, they don't

---

[13] Maybe there is plenty of independent variation in this world. But what about simple worlds, such as a world containing nothing but two hard atoms which collide with each other? My answer (and Hausman's, 1998, pp. 67-8) is that such worlds would lack causal direction. This thought might help readers to understand *why* effects should always have independent sources of variation. At first sight it might seem a lucky freak that each effect should have some such independent source for each of its causes. But it's not so odd, if that is what *makes* this effect an effect of that cause. The lucky thing is only that the world should display sufficiently complex patterns of probabilistic dependence and independence. No extra design is needed to link up these patterns appropriately with causes and effects—this linkage simply falls out of metaphysical essence of causal direction.

always do so, and when they don't the link between causes and correlations is broken.

I am not convinced these difficulties are insuperable. In the next two subsections I shall consider the first kind of query—are there correlations which lack causal significance? After that, I shall consider the converse question—are there causal connections which fail to manifest themselves as correlations?

6.1 Bread Prices and Venice Water Levels

Aren't there plenty of obvious correlations which signify no causal connection? Isn't there is a good correlation, for instance, between bread prices in Britain and water levels in Venice? (They have both been steadily rising since records began.)

One ploy here would be to query whether such cases are genuinely lawlike, as opposed to sampling artifacts (cf. footnote 3). But I shall not push this line. There may be something odd about the correlation between bread prices and water levels, but I am prepared to accept that it is a genuinely projectible pattern which we can expect to hold up in future cases.

Instead I would like to discount this correlation on the grounds that its instances are inappropriately related. In the normal case, correlations are calculated from the paired values displayed when two variables are spatiotemporally co-instantiated—this is why these paired values are candidates for causal relationships. (My school and my exam results; your school and your exam results; and so on.) By contrast, the values displayed in different instances will normally bear no specified spatiotemporal relationship to each other, so will not raise any questions of causal influence. (If my school is affecting your exam results, we need a more complicated model, which respects the requirement that only spatiotemporally co-instantiated values of variables are candidates for causal relationships. For example, we could take the instances of the correlation to be pairs of appropriately spatiotemporally related people.)

The bread-Venice example doesn't conform to these requirements. The bread prices at one time affect those at other times, as do the Venice water levels. Because of this, the correlation here doesn't represent a connection between bread price and water levels which is due solely to the co-instantiation of these variables. Rather it reflects the fact that temporally earlier bread prices influence later ones, and temporally earlier water levels influence later ones. Bread prices predict Venice water levels, and vice versa, only because the paired values have independently undergone similar causal histories. From now on, accordingly, I shall take it as read that we are dealing with correlations whose different instances bear no specified spatiotemporal relationships to each other, and so can't have been generated in this way.

Note in this connection how the bread-Venice example will display no co-variation between bread and water levels beyond the correlation that is already implied by each times series taken separately. That is, there isn't any tendency for the two series to peak or fall in tandem. If we subtract that correlation between bread and water levels that can be attributed to the temporal succession of instances, then we are left with no correlation at all. (If there were such a residual correlation, then that would indeed point to a common cause.)

6.2 EPR Correlations

The other obvious problem for the claim that all correlations have causal significance is quantum mechanics. In particular, EPR cases certainly involve correlations, but it seems unlikely that these reflect causal connections. Apart from anything else, the correlated events can be spacelike separated, which means that any causal influence between them would be in tension with special relativity.

In fact, Hausman's methods deal with this kind of case quite naturally (Hausman 1998, sects 12.6-7).[14] The results on the different wings of an EPR experiment *both* lack independent sources of variation. Anything correlated with one is correlated with the other, and vice versa. So Hausman's principle (II) implies that neither is an effect of the other, nor do they have a common cause.

Accordingly, Hausman defines a notion of 'mutual dependence': two events are mutually dependent if everything correlated with one is correlated to the other. And then he modifies accordingly:

(I) A and B are correlated *if and only if* either A causes B, or B causes A, or they have a common cause, or they are mutually dependent.

Since the reductive principle (III) already required that we have cause-effect relationships only if one end of the correlation has an independent source of variation, this reductive principle can stand as before.

6.3 Failures of Faithfulness

Let me now consider the converse problem facing reductionism, the possibility that there are causal connections which fail to manifest themselves as correlations. In the literature these have come to be known as 'failures of

---

[14] Given Hausman's non-probabilistic programme, his worry isn't that EPR cases yield non-causal *correlations,* but that they yield non-causal nomic connections. Still, his proposed solution is equally available to my probabilistic programme.

faithfulness'—the correlations we observe are not faithful to the underlying causal structure.

Now, it seems all too possible that there should be such cases. To repeat an example I have used previously, suppose that drinking cola (C) both stimulates people to exercise more (E), but also causes them to put on weight (W). And suppose further that exercise E independently has a negative influence on weight increase W, to just the extent required to cancel out the direct positive influence of C, and leave us with an overall zero correlation between cola C and weight increase W.

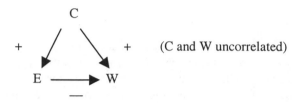

Here we have a causal connection between C and W, but no corresponding correlation. Indeed if we were to look only at the correlations here, we would get the impression that C and W were both causes of E, but were themselves causally unconnected. (After all, they are themselves uncorrelated, but both correlated with E.)

Now clearly this kind of case is *unlikely*. It requires two causal influences to cancel out precisely, and so to leave us with no correlation at all. It thus involves a kind of freak, which can perhaps be ignored for practical purposes.[15] But this freakiness is no help to the reductionist programme. For a reductionist has to say that such failures of faithfulness are not just unlikely, but metaphysically *impossible*. And this seems just wrong. There seems no principled barrier whatsoever to two causal influences canceling out exactly and leaving us with a zero correlation, as in the above example. Maybe this would be a freak, but as long as it *can* happen, reductionism is in trouble.

Both Irzik and Hausman regard this as the Achilles' heel of the reductionist programme (Irzik 1996, sect 6; Hausman 1998, sect 10.4). From their point of view, such examples show that causal connections lie deeper in reality that mere correlations. Non-reductionists can allow that causal

---

[15] It would be foolish, however, to ignore this danger too readily when our information about correlations comes from small samples. Remember (footnote 1) that survey research involves two stages: first, inferences from samples to correlations; second, from correlations to causes. Even if a freakish canceling out of precise causal influences is needed to undermine the second stage, *approximate* canceling out of causal influences can easily mess up the first stage, by preventing a genuine causal correlation from registering as statistically significant.

connections *normally* manifest themselves in correlations. Perhaps they can even think of them as dispositions to generate correlations when combined in various ways. But they won't equate the causal connections with actually manifested correlations, precisely because they will take it to be entirely possible that certain combinations of underlying causal dispositions will generate the 'wrong' correlations, as in failures of faithfulness.

In earlier work (1993a) I suggested that this difficulty might be avoided by attending to yet further variables. For instance, if in the above example we could find some Z which is correlated with Weight but not with Exercise, then this would argue that Weight cannot be a cause of Exercise after all, despite initial indications. And perhaps a wider network of variables will suffice to pin down the correct causal structure uniquely.

Hausman has pointed out, however, that this strategy is unlikely to serve the reductionist's purposes. For, even if such a wider network can yield additional fixes on the underlying causal structure, the original failure of faithfulness will not be removed. If the unconditional correlation between Cola and Weight is zero, then it will still be zero after I have examined various further correlations. So whatever principles I might be using to identify the real causal structure, they cannot include the basic reductionist premise (I) which says that absence of correlation means absence of causal connection. *This* principle is falsified by the failure of faithfulness. At best I must be using something like "Whenever different aspects of a correlational structure imply inconsistent causal conclusions, postulate as few failures of faithfulness as possible to resolve the inconsistencies". (The idea here would be that if I *did* continue denying a causal connection between Cola and Weight, in the face of extra correlational evidence, I would need to postulate even more failures of faithfulness to explain away that extra evidence.)

While this principle of charity about fidelity seems a sensible enough methodological maxim, I agree with Hausman that it is an unlikely basis for a metaphysical reduction of causation. So instead I would like to adopt a different strategy, and appeal to the metaphysical picture developed earlier in the paper to answer the challenge.

The idea here would be to go finer rather than broader. Instead of looking at wider frameworks of variables, including more distal causes and effects of such initially troublesome trios as Cola, Exercise and Weight, we could switch the focus to a more microscopic level, and include factors which mediate causally between these variables (cf. Hausman 1998, p. 221)[16]. If the arguments in the first half of this paper are correct, there is

---

[16] Hausman also makes the point (1998, p. 215) that a pseudo-indeterminist like myself is in danger of having correlations go degenerate if I include all relevant microscopic variables,

every reason to suppose there are many such variables, indeed enough to restore pseudo-indeterminism, and moreover that they will satisfy probabilistic independence requirements on causes. The hope, then, would be that at this level there will be no failures of faithfulness, and that the reductive principles derived from Hausman will suffice to fix the correct causal order.

Still, it is not enough that there should simply *happen* to be no failures of faithfulness at such more microscopic levels. Reductionism requires that there *couldn't* be any such failures. And it might seem that I still have no argument here. Won't it still make perfectly good metaphysical sense that there should be some fortuitous canceling of causal influences at the lower levels, thus violating faithfulness once more?

However, there is no reason for the reductionist to accept that such failures of faithfulness can go all the way down. This might make initial conceptual sense, but the reductionist can insist that, since the causal order is metaphysically fixed by probabilistic patterns at microscopic orders, there is simply no metaphysical possibility left of such a microscopic pattern painting a false causal picture.

From my reductionist perspective, to think that there is always a deeper level of *causal* influences underlying the probabilities is to mistake a difference of levels for a difference in metaphysical kind. It is true that, if we start at a macroscopic level, then there is a deeper level of reality underlying the probabilities at that level, which can discredit those probabilities as causally misleading. But this isn't because that deeper level consists of something non-probabilistic and causally sui generis. Rather it is simply that the probabilities at the micro-levels trump those at the macro-levels as a means of fixing causal order.

Compare the response of the beginning student when exposed to the textbook Humean view that causation is nothing but constant conjunction. "That can't be right, because it leaves out the continuous *mechanisms* connecting causes and effects." One can see what the student means, but this thought on its own is no threat to the Humean position. For, as the teacher will point out, the interesting question is whether the more fine-grained causal links which make up the 'mechanisms' are themselves cemented by anything more than constant conjunction.

---

since the probabilities will then all go to zero or one. But this is too quick: even if determinism means that some (total) antecedents fix zero and one for consequents, less-than-total antecedents can still fix intermediate probabilities. True, this reductionist response does implicitly appeal to the 'naturalness' of certain kinds. But Hausman himself needs natural kinds (pp. 66-8). Moreover, the reductionist programme is likely to need them elsewhere as well, to respond to the kind of challenges raised by Arntzenius (1990) and Price (1996).

Similarly in the present context. You may feel intuitively that there must real *causal mechanisms* behind the probabilities linking causes and effects. But this feeling in itself is no serious threat to the reductionist position, since reductionists can simply respond that the links in such mechanisms are nothing but probabilistic patterns at a more fine-grained level.

If reductionists take this line, they owe some explanation of which 'levels' are to count in fixing causal order. They must dismiss probabilities at 'higher' levels as not themselves being constitutive of causal order, otherwise they will be left with no answer to failures of faithfulness at that level. But then which levels do count? Gesturing at an ordering into 'higher' and 'lower' levels does not really serve, since it fails to tell us which probabilities are constitutive of causal relationships and which are not.

The natural reductionist answer is that only the *lowest* level counts. Causal relationships at higher levels are fixed by those at the lowest level. Patterns of correlation can thus be misleading about causal structure at any higher level. But at the bottom level there is no metaphysical room for such failures of faithfulness, since there the causal order is simply constituted by the correlational order.

What if there is no lowest level, if there is no limit to how fine we can cut up our mechanisms? Then reductionists can adopt a limiting procedure. Provided there is an ordering of levels into more or less fine-cut, they can say that causal order is fixed once we reach a level where no lower level's correlations overturn that order.

Unsympathetic readers are likely to feel that the reductionist programme is now resting on speculation. To which I readily concede that there is much that is unclear here. In particular, I would like a better understanding of the interface between purely quantum mechanical situations, where causal order goes fuzzy in ways touched on earlier, and the 'pseudo-indeterministic' world, where events become determinate enough to fall into patterns constituting causal order.

At the same time, I see no need to apologise for my metaphysical commitments. As I said at the beginning of this paper, a reduction of causation to probabilities would be a fine thing, if it were possible. For it would show us how there can be causal direction in a world whose fundamental laws are symmetrical in time. Given this, there seems to me no need for every plank in the reductionist programme to be nailed down firmly. If we can develop a cogent metaphysical picture of the sources of causal aymmetry, and if this picture can be supported by general considerations, of the kind offered in the first half of this paper, then I would say we have a good theory of causal asymmetry.

References

Arntzenius, F. (1990) Physics and common causes. *Synthese,* 82

Beebee, H. and Papineau, D. (1997) Probability as a guide to life. *Journal of Philosophy,* 94, 217-243.

Eells, E. (1991) *Probabilistic Causality.* Cambridge: Cambridge University Press.

Hausman, D. (1998) *Causal Asymmetries.* Cambridge: Cambridge University Press.

Irzik, G. (1996) Can causes be reduced to correlations? *British Journal for the Philosophy of Science,* 47, 249-270.

Papineau, D. (1985a) Probabilities and causes. *Journal of Philosophy,* 82.

Papineau, D. (1985b) Causal asymmetry. *British Journal for the Philosophy of Science,* 36, 273-289.

Papineau, D. (1989) Pure, mixed and spurious probabilities and their significance for a reductionist theory of causation. In P. Kitcher and W. Salmon (Eds.) *Scientific Explanation: Minnesota Studies in the Philosophy of Science vol XIII.* Minneapolis: University of Minnesota Press, 307-348.

Papineau, D. (1993a) Can we reduce causal direction to probabilities? In D. Hull, M. Forbes and K. Okruhlik (Eds.), PSA 1992 vol 2. East Lansing: Philosophy of Science Association, 238-252.

Papineau, D. (1993b) The virtues of randomization. *British Journal for the Philosophy of Science*, 44, 437-450.

Papineau, D. (forthcoming) Causation as a guide to life.

Pearl, J. and Verma, T. (1994) A theory of inferred causation. In D. Prawitz, B. Skyrms and D. Westerståhl (Eds), *Logic, Methodology and Philosophy of Science IX* Amsterdam: Elsevier, 789-811.

Price, H. (1996) *Time's Arrow and Archimedes' Point.* Oxford: Oxford University Press

Reichenbach, H. (1956) *The Direction of Time.* Berkeley: University of California Press.

Spirtes, P, Glymour, C. and Scheines, R. (1993) *Causation, Prediction and Search.* New York: Springer-Verlag.

Suppes, P. (1970) *A Probabilistic Theory of Causality.* Amsterdam: North Holland.

3

Probabilistic Causality, Direct Causes and Counterfactual Dependence[*]

JAMES WOODWARD

California Institute of Technology

1 Introduction

A great deal of recent philosophical work on causation largely falls in two
major traditions. On the one hand, there is the tradition of probabilistic theories of causality inaugurated by Suppes (1970). Suppes hoped to reduce
causal claims to claims about probabilities. More recent work in this tradition
eschews the goal of a complete reduction but still hopes to find systematic
relationships between causal claims and claims about probabilities. For example, a standard suggestion (Cartwright 1983) about that connection is that
causes must raise the probability of their effects across all causally homogenous background contexts or given all possible combinations of other factors
that are causally relevant to the effect. Theories of this sort are generally intended as theories of so-called type causation: that is they are intended to
capture causal claims that relate types of events or properties such as "im-
pacts of rocks cause windows to break" and "smoking causes lung cancer".
With a few exceptions (e. g., Eells 1991), they are generally not intended to
be accounts of so-called token causation—that is, causal claims that relate

[*] Thanks to Chris Hitchcock for a number of helpful discussions.

Stochastic Causality.
Maria Carla Galavotti, Patrick Suppes and Domenico Costantini (eds.).
Copyright © 2001, CSLI Publications.

individual events such as "the impact of the ball thrown by Billy on January 12, 2000 at 3pm caused Smith's window to shatter".

One of the main alternatives to such theories in philosophy is the counterfactual approach to causation developed by David Lewis (Lewis, 1973, 1979) and his students. This approach is also reductionist in intent, but in contrast to probabilistic theories, the aim is to reduce causal claims to claims about counterfactual dependence, where the latter can be understood in a way that does not presuppose causal ideas. Moreover, in contrast to probabilistic theories, Lewis's theory is not intended as an account of type causation but rather as an account of token causation. Lewis's theory does succeed in capturing our common sensical judgments about token causal relationships in a range of (although by no means all) cases. However the theory requires a semantics for counterfactuals that is prima-facie quite mysterious—for example, we are often required to employ counterfactuals the antecedents of which are made true by miracles. Many writers have argued that such counterfactuals play no legitimate role in scientific practice.

What is the relationship between these philosophical theories and the treatments of causal claims one finds in discussions of experimental design and in disciplines like econometrics and epidemiology—treatments that employ, for example, the apparatus of structural equations and/or directed graphs? These disciplines differ along many different dimensions and there is no single generally accepted label for the work on causal inference and the representation of causal relationships one finds in them. To avoid cumbersome repetition, I will call them the causal modeling disciplines and will speak (in a way that obviously involves considerable idealization) of "the" causal modeling conception of causation. Causal inference in the causal modeling disciplines is usually based, at least in part, on statistical evidence and, in part for this reason, a great deal of work in these disciplines focuses on the relationship between causal claims and probabilistic relationships. This fact has led many philosophers to suppose that the causal modeling notion of causation must be something like the notion that probabilistic treatments of causation attempt to capture. Indeed, the assumption that this is the case has been one of the main motivations for the development of probabilistic theories.

In this essay, I will compare the treatment of causation assumed in the causal modeling disciplines with the probabilistic and counterfactual theories developed by philosophers. I will suggest that, contrary to what many philosophers have supposed, the causal modeling notion is in many respects closer to the counterfactual approach assumed in the Lewisian tradition than to probabilistic accounts of causality. In fact, the causal modeling treatment of causation clarifies and explains some of the puzzling features of Lewis's account of counterfactuals. Moreover, the causal modeling treatment yields a notion of causation that, provided one is willing to make certain assumptions, does have systematic connections with facts about probabilities. However, these connections are considerably looser than those defended in standard formulations of probabilistic theories—so loose that the phrase "probabilistic theory of causation" seems a misnomer. In general, the causal modeling conception of causation brings together work in both the counterfactual and probabilistic causality traditions, capturing what is plausible in both.

As a point of departure, let me begin with an observation whose significance is insufficiently appreciated by many philosophers working in the probabilistic causality tradition. This is that the standard treatments of causal relationships one finds in the causal modeling disciplines employs two distinct kinds of resources or representations, both of which work together in problems of causal inference. First, as in probabilistic theories of causation it is assumed that we have a probability distribution **P** over some set of variables V whose causal relationships one is interested in investigating. Second, over and above this, one makes use of some additional device **G** to represent causal relationships among the variables in **V**. **G** is typically either system of equations or a directed graph. We may thus think of a causal model as an ordered triple of the form <**V, P, G**>. As we will see, if we are to adequately represent the connection between casual relationships and probabilities, *both* **P** and **G** are required—neither cannot be reduced to or replaced by the other. Roughly speaking the role of the directed graphs or structural equations is to represent information about patterns of counterfactual dependence among variables; more specifically, it is to tells us what would happen to the values of some variables under changes of a special sort involving what I will call *interventions*. (These are idealized experimental manipulations—see Section 4) in the values of other variables. The probability distribution **P**, by contrast, does not convey modal or counterfactual information of this sort. Instead, it conveys information about the actual distribution of values of variables. As we will see below, we may think of directed graphs and structural equations not as summarizing information about any particular probability distribution but rather as telling us various different distributions are connected to one another—in particular how such distributions (or certain features of them) will change under interventions or combinations of interventions. Seen from this perspective, a major problem with the probabilistic theories found in much of the philosophical literature is that they attempt to provide an account of causation by relying too heavily on just one kind of resource—the probability distribution **P**. In fact the full resources of the graphical or equational representation are required if one is to do justice to the notion of causation.

2   Directed Graphs and Equations

I begin with some brief remarks about the use of directed graphs and systems of equations to represent causal relationships. Let us assume that the causal claims that we are interested in modeling relate variables where variables represent properties or magnitudes that (as the name implies) are capable of taking more than one value[1]. Such claims will involve type-level rather than token level relationships. The familiar examples of so-called property or type causation discussed in the philosophical literature may be understood as relationships between two-valued or binary variables, with the variables in question taking one or another of two values, depending on whether the properties in question are instantiated or occur. Thus the claim that ingestion of aspirin causes recovery from headache may be understood as asserting a relationship of some kind between a variable A, representing whether or not aspirin is ingested, and a variable H representing whether or not relief from headache occurs. Of course variables need not be two-valued; they may also assume many values or be continuous.

A directed graph is an ordered pair $<V, E>$ where V is a set of vertices which serve as the variables representing the *relata* of casual relationships and E a set of directed edges connecting these vertices. A directed edge from vertex or variable X to vertex or variable Y means that X *directly causes Y*. For now I will largely rely on the reader's intuitive understanding of this notion; it is discussed in more detail in Section 5 below. However, the basic idea is that X is a direct cause of Y if and only if there is a possible intervention (experimental manipulation) on X that would change the value of Y (or the probability distribution of Y) when all other variables in the system of interest are held fixed at some set of values in a way that is independent of the change in X. Put more simply, drawing an arrow from X to Y means that there is *some* change in the value of X that will change the value of Y (or the probability distribution of Y), given *some* set of values for the other variables. I assume that if X is a direct cause of Y, then X is a cause of Y, but that the converse of this claim is false. A sequence of variables $\{V_1, \ldots V_n\}$ is a *directed path* or *route* from V_1 to V_n if and only if for all i ($1 \leq i \leq n$) there is a directed edge from V_i to V_{i+1}. Y is a *descendant* of X if and only if there is a directed path from X to Y. If Y is a descendant of X, then X is an *ancestor* of Y. The direct causes of X are also said to be the *parents* of X[2]. As we will see below,

---

[1] To avoid cumbersome terminology, I will often use the word "variable" to refer both to the properties etc. that serve as *relata* in causal relationships and to the symbols that represent such properties. This conflation is, I believe, harmless.

[2] See Spirtes, Glymour and Scheines, 1993 for a more detailed discussion of the use of directed graphs to represent causal relationships.

a necessary but not a sufficient condition for X to be a cause of Y is that X be an ancestor of Y.

As an illustration of the use of directed graphs to represent causal relationships, suppose that A is a variable measuring atmospheric pressure, B a variable representing the reading of a particular thermometer and S a variable representing the occurrence or non-occurrence of a storm Then we may represent the claim that A is a direct common cause of B and S and that B does not cause S and S does not cause B by means of the following diagram.

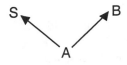

Figure 2.1

Causal relationships may also be represented by means of systems of equations. When underlying causal relationships are deterministic each endogenous variable Y (that is, each variable that represents an effect) is written as a function of all and only those variables that are its known or measured direct causes plus a so-called error term which represents the combined influence of all of the other direct causes of the endogenous variable. The presence of the error term makes possible conditional probabilities involving the measured variables that are strictly between zero and one. There is one equation for each endogenous variable. For example, if variables X_1, ..., X_m are all of the known direct causes of Y, then Y may be written as (1.1) $Y = F_Y$ $(X_1,...,X_m) + U$. Analogous remarks apply to the indeterministic case, with the relevant equations specifying how the probability distribution of Y will change under manipulation of the right side variables representing direct causes in each equation.

What is the relationship between the representation of causal relationships by means of systems of equations and their representation by means of directed graphs? As we have seen, when we draw a directed graph with arrows from X_1..., X_m into Y, we convey the information that Y is some function of X_1..X_m and that all of the variables X_1, .., X_m are essential in the sense that for each such variable X_i , there is some combination of values of the others such that changing X_i will change Y. However, the graph does not further specify what this function is. It does not tell us *which* changes in X will change Y or by how much or for which values of other variables. By contrast, when we explicitly specify the function or equation relating Y to its direct causes (e. g. $Y = 3X_1 + 4X_2$), we convey more information than if we merely draw a graph with arrows from X_1 and X_2 directed into Y. Unlike the directed graph, the explicit form of the equation specifies exactly how changing X_1 and X_2 will change Y. By contrast, the corresponding directed graph says simply that there is some change in X_1 (X_2) that, given some value of X_2 (X_1), will change Y.

3   Causation, Manipulation and Counterfactuals

I said above that directed graphs and systems of equations represent counter-factual claims about how changing the values of certain variables will change the values (or the probability distribution of the values) of others. Let me now try to be more precise about this idea and its underlying motivation. The notion of causation assumed in directed graphs and systems of equations is a *manipulability* conception of causation. The underlying idea is that causal relationships are just those relationships that are potentially usable for pur-poses of manipulation and control in the sense that if X is a cause of Y then if it were possible to change or wiggle the value of X in the right way and in the right circumstances, this would be a way of wiggling or changing the value of Y. Manipulability theories are thus a subspecies of counterfactual theories of causation; they are theories according to which the right counterfactuals for understanding causal claims are counterfactuals that have to do with what would happen under hypothetical manipulations.

Manipulability accounts of causation have been unpopular in contempo-rary philosophy; they are commonly criticized as both unilluminatingly cir-cular and as leading to an unacceptably anthropomorphic or subjective no-tion of causation in the sense that they seem to restrict true or meaningful causal claims to those contexts in which manipulation by human being is possible (see, e.g., Hausman 1998). I have argued elsewhere (Woodward forthcoming) that while these criticisms are indeed apt when applied to the standard formulations of the manipulability theory one finds in the philo-sophical literature (such as von Wright 1971; Price 1991), there is natural way of developing an alternative version of manipulability theory that avoids such criticisms. As we will see (Section 4), the key to formulating an accept-able version of the manipuability theory is finding the right characterization of the notion of an intervention: an intervention on X with respect to Y can be characterized in a way that makes no reference to human beings or their ac-tivities (thus avoiding the anthropocentrism of traditional versions of the manipulability theory) and also in a way that makes no reference to the exis-tence or non-existence of a causal relationship between X and Y (thus avoid-ing the vicious circularity that infects traditional versions).

But while it is possible in this way to formulate a version of the manipu-lability theory that avoids the standard criticisms, a natural (and deeper ques-tion) is why we should bother to do this. Why suppose that we can clarify or explain anything about causal relationships by thinking about them within the framework of a manipulability theory? For reasons of space, I will confine myself to a few brief observations. First, researchers within the causal modeling disciplines tell us again and again that what they *mean* by causal relationships are those relationships that are exploitable for purposes of manipulation and control[3]. While it is of course possible that such pronouncements have no relation to the conception of causation assumed in the actual practice of these disciplines, this is a good prima-facie reason for taking the manipulability conception seriously.

Second, manipulability theories of causation provide a natural and attractive account of the underlying point or rationale of our practice of distinguishing between causal and non-causal relationships: if X and Y are correlated and if manipulation of X is possible, there are obvious practical advantages to knowing whether or not the relationship between X and Y is such that manipulating X can change Y. Moreover, at least in many contexts, it also seems clear that it is exactly this distinction that is at issue when we worry about whether a relationship is causal. Consider the well documented correlation between superior scholastic performance and attendance at private schools. Does this reflect a (3.1) causal connection between such attendance and performance or (3.2) is it rather the case that private school attendance per se has no effect on performance and that the correlation arises entirely from the fact that the very factors that lead to enrollment in private schools (e.g., affluent parents who are concerned about their children's education) also cause superior performance? Parents and educational researchers care about the answer to these questions exactly because they want to know whether they can manipulate performance by enrolling students in private schools—it is possible in principle to do this if (3.1) is correct but not if (3.2) is. More generally, human beings are often in the position of observing a correlation between X and Y and wondering whether this correlation reflects a relationship that will allow them to change Y by manipulating X or whether instead the observed correlation between X and Y will disappear under manipulation of X. According to a manipulability theory our notion of causation developed not as the result of an impulse to engage in dubious metaphysics or to project certain of our psychological states onto the world but rather to mark this practically important distinction. On this view, directed graphs and

---

[3] Illustrations are readily found in a variety of texts on experimental design and econometrics. A representative quotation from Cook and Campbell's highly influential (1979) is: *The paradigmatic assertion in causal relationships is that manipulation of a cause will result in the manipulation of an effect.* Causation implies that by varying one factor I can make another vary (1979, p. 36, emphasis in original).

systems of equations have a similar motivation; they reflect our concern to distinguish manipulation supporting relationships from mere correlations.

Like the notion of causation itself, counterfactuals have often been regarded with suspicion by empiricists. It is frequently suggested that they lack a clear meaning or that their truth conditions are so vague and context-dependent that they are not suitable for understanding or elucidating any notion (of causation or anything else) that might be of scientific interest. A famous example of Quine's illustrates the worry. Consider the counterfactual(s) (3.3) "If Julius Caesar had been in charge of U. N. forces during the Korean war, then he would have used (a) nuclear weapons or (b) catapults". It is hard to see on what basis one could decide whether the counterfactual (3.3) with (a) as consequent or the counterfactual (3.3) with (b) as consequent (or neither) is correct. A manipulability framework for understanding causation helps to address this worry. It suggests that the appropriate counterfactuals for elucidating causal claims are not just any counterfactuals but rather counterfactuals of a very special sort: those that have to do with the outcomes of hypothetical manipulations or experiments. It does seem plausible that counterfactuals that we do not know how to interpret as (or associate with) claims about the outcomes of well-defined manipulations will often be claims that lack a clear meaning or truth value. For example, (3.3a) and (3.3b) seem unclear for just this reason. It isn't just that we lack the technological means to carry out an experimental manipulation in which Caesar is placed in charge of the U. N. forces. The more fundamental problem is that we have no clear conception of what would be involved in carrying out such an experiment.

By contrast, a similar sort of skepticism about counterfactuals that are interpretable as claims about the outcomes of hypothetical (but otherwise well specified) experimental manipulations is much harder to sustain. Consider an experiment in which a large group of people suffering from a disease are randomly divided into a treatment and a control group with the former receiving some drug that is withheld from the latter. As it turns out, the incidence of recovery is much higher in the former than in the latter. Provided that the right sort of experimental controls have been followed, it is very natural to think of this experiment as providing good evidence for the truth of counterfactuals like the following: (3.4) "If those in the control group had received the drug, the incidence (or expected incidence) of recovery in that group would have been much higher." Indeed it is very plausible that it is precisely because the experimenters want to determine the truth value of counterfactuals like (3.4) that they conduct the experiment. Of course, the researchers may be mistaken in the conclusion they draw about the truth value of (3.4) but this does not distinguish (3.4) from any other empirical knowledge claim. The claims that (3.4) lacks a determinate meaning or truth value or is untestable in principle are, I suggest, much less plausible than the corresponding claims about (3.3)[4].

In contrast to counterfactuals like (3.3), counterfactuals like (3.4) play a central role both in practical deliberation and in experimental practice in science. We need to understand how such counterfactuals can be tested and evaluated but they should not be dismissed as meaningless or unscientific.

4   Interventions

I suggested above that one of the key elements in formulating a defensible version of a manipulability theory is the notion of an intervention. Heuristically (but only heuristically), one may think of an intervention on X with respect to Y as the sort of manipulation that might be carried out in an ideal experiment for the purpose of determining whether X causes Y. The basic idea may be illustrated by reference to the ABS system in Figure 2.1. It is clear that there are ways of changing B that will be associated with a corresponding change in S even though B does not cause S. For example, if we change B by changing A, or by means of some causal process that is perfectly correlated with changes in A, then S will also change, but this would not establish that B causes S. Plainly, an experiment in which B is manipulated in this way is a badly designed experiment for the purposes of determining

---

4 These remarks raise a natural question. Suppose that we grant that counterfactuals like (3.4) that can be tested experimentally have truth values and hence that the causal claims associated with them have truth values as well. What about causal claims and associated counterfactuals for which the relevant experimental manipulations are specifiable or well-defined but cannot actually be carried out, because of technological or other sorts of limitations? For example, consider the causal claim that (3.5) the position of the moon causally influences the tides and the associated counterfactual claim that (3.6) if the radius of the moon's orbit were to be changed as a result of an intervention, the motions of the tides would have been different. Assuming that the causal claim (3.5) is true, on a manipulability theory some associated counterfactual like (3.6) must be true as well. But why suppose that (3.6) has a definite truth value if, as is clearly the case, the associated manipulation cannot actually be carried out? While the matter deserves a more detailed discussion than I can give it here, the short answer is that once it is accepted that (a) counterfactuals have truth values when their antecedents refer to experiments that can be carried out, it is hard to avoid the view that (b) counterfactuals for which the associated experiments are well-defined but cannot be carried out also have truth values. The reason for this is that even for counterfactuals satisfying (a), it is not the actual carrying out of the associated experimental manipulations that endows them with definite truth values. Rather, such counterfactuals possess definite truth values independently of whether the relevant experimental manipulations are carried out. The experimental manipulations are a way of discovering what the truth values of the counterfactuals are; they do not somehow create those truth values. Similarly for counterfactuals satisfying (b)—if the manipulations specified in their antecedents cannot, as a practical matter, be carried out, this shows only that their truth or falsity cannot (at present) be directly determined by experimentation, not that they lack truth values.

whether B causes S. Similarly, an experiment in which the process that changes B also directly changes S would be badly designed for this purpose.

By contrast, consider the following experiment. We employ a random number generator which is causally independent of A and, depending just on the output of this device, repeatedly physically fix the barometer reading at different values by moving the dial to either a high or low reading and driving a nail through it. If it is really true that B does not cause S, then we expect that the changes in B produced by such interventions will no longer be associated with changes in S. If, on the contrary, S continues to be correlated with S under such interventions on B, this would be strong prima-facie evidence that B does cause S. In contrast to the previous experiments, an experiment of this sort would be a well designed experiment for the purposes of determining whether B causes S. The notion of an intervention is meant to capture the contrast between these two kinds of experiments: the second sort of experiment involves an intervention on B with respect to S while the first does not.

This reference to an "ideal experiment" naturally suggests an activity carried out by human beings. However, as I suggested above, the notion of an intervention can be given a completely nonanthropomorphic characterization, that makes no reference to human beings or their activities. I will not try to present the full details of this characterization here, but instead refer the reader to the characterization in Woodward (2000)[5]. Informally, however, we may think of an intervention I on X with respect to Y as an exogenous causal process that changes X in such a way and under conditions such that if any change occurs in Y, it occurs only in virtue of Ys relationship to X and not in any other way. Making such a characterization precise requires reference to the causal relationships between I and various other causes of Y and to the causal relationship between I and Y (for example, I must not be a direct cause of Y) but it does not require reference to the presence or absence of a causal relationship between X and Y. Thus, such a characterization will not be viciously circular in the sense that to know whether an intervention has been carried out on X with respect to Y, one must already know whether X causes Y.

The sense in which interventions involve *exogenous* changes in the variable intervened on is illustrated by the above example. When an intervention occurs on B, the value of B is determined entirely by the intervention, in a way that is (causally and probabilistically) independent of the value of A which was what previously determined the value of B. In this sense the intervention breaks or disrupts the previously existing endogenous causal relationship between A and B. If we represent such an intervention on B by

---

[5] For additional and broadly similar characterizations of the notion of an intervention see Spirtes, Glymour and Scheines (1993), Woodward (1997), Hausman (1998), and Pearl (2000).

drawing an arrow from an intervention variable I to B, then the result of the intervention will be to replace the structure in Figure 2.1 with the structure in Figure 4.1:

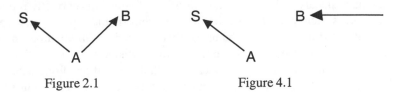

Figure 2.1 Figure 4.1

This illustrates the so-called "arrow-breaking" conception of interventions: an intervention on X breaks all arrows directed into X while preserving all other arrows in the graph, including those directed out of X. The breaking of arrows directed into X captures the idea that the value of X is now determined exogenously, entirely by the intervention variable I and that those variables that influenced the value of X prior to the intervention no longer do so[6].

This idea about how to represent interventions graphically is closely tied to an idea about the impact of interventions on systems of equations that Dan Hausman and I (Hausman and Woodward 1999) have elsewhere called "modularity". We may represent the structure in Figure 2.1 by means of the following two equations

(4.1) $B = aA$

(4.2) $S = bA$

An intervention on B will then correspond to replacing equation (4.1) with a different equation (4.3), $B = I$, specifying that the value of B is no longer determined by S but is instead set entirely by the value of the intervention variable I. Just as we assume, when we employ the arrow-breaking conception of interventions, that it is possible to carry out an intervention on B that leaves the arrow from A to S undisturbed, we also assume that when a system of equations like (4.1- 4.2) correctly represents some causal structure, it will be possible to carry out this operation of replacing one equation in the system (in this case, (4.1) with (4.3)) while leaving the other equations in the system (4.2) undisturbed. When a system of equations has this feature (that is, when one may disrupt or replace any one of the equations by means of an intervention on the dependent variable in that equation, without disrupting the other equations), I will say that the system is *modular* or equation-invariant. Within a probabilistic framework, modularity corresponds to the

---

[6] For additional discussion of the arrow-breaking interpretation, see Pearl (1995, 2000), Spirtes, Glymour and Scheines (1993) and Hausman and Woodward (1999).

following requirement: $Pr(X/Parents(X).Set\ Y) = Pr(X/Parents\ (X)$ for all Y distinct from X, where Set Y means that the value of Y has been set by an intervention.

One natural way of motivating the idea that systems of equations should be modular appeals to the idea that if a system of equations correctly represents causal structure, each equation in the system should represent the operation of a distinct causal mechanism. If we make the additional plausible assumption that a necessary condition for two mechanisms to be distinct is that it be possible (in principle) to interfere with the operation of one without interfering with the operation of the other and vice-versa , we have a justification for requiring that systems of equations that correctly represent casual structure should be modular. For example, a natural justification for supposing that we may replace (4.1) with (4.3) without altering (4.2) is that the mechanism by which A affects B is distinct from the mechanism by which A affects S. In what follows, I will assume that the systems of equations with which we are dealing are modular (and correlatively that graphical representations satisfy the arrow-breaking interpretation of interventions). For example, the definition of direct causation (**DC**) given below assumes modularity.

It also will be important to our subsequent discussion to understand that *intervening* to set the value of a variable to some value is conceptually quite different from *conditioning* on the value of that variable. (cf. Meek and Glymour 1994; Pearl 2000). Following a proposal due to Pearl (1995) let us suppose that the values of variable X that are set by interventions can be represented as the values of a new random variable, *set X*. (This will be a reasonable assumption when, as in the example above, the values of this variable are determined by a randomizing device). Then it will not in general be true that $Pr\ (Y/X) = Pr\ (Y/set\ X)$. In the above example, S and B are correlated and in fact $Pr(S/B) > Pr(S)$. However, assuming that B does not cause S, we would not expect S and *set B* to be correlated. Instead we would expect that $Pr\ (S/set\ B) = Pr\ (S)$.

The reason why intervening is different from conditioning is unmysterious. When we condition on a variable, we assume that whatever causal structure generates the values of that variable is left intact, so that the values in question continue to be generated by whatever endogenous causal factors are at work in that system. Thus when we condition on B in the above example, we assume that the values of B continue to be generated by A, in which case they will be correlated with the values of S, which are also generated by A. By contrast, as the above example illustrates, if a variable is endogenous, then intervening on it alters the causal structure of the system in which it figures—giving it a new exogenous causal history. Unless the variable intervened on is exogenous, the intervention will disrupt the previously existing pattern of correlations in the system, leading to a new set of probability relationships. We see this in the example under discussion, in which B and S are correlated when there is no intervention on B, but are uncorrelated in the new structure that results when there is an intervention on B. (The graph in Figure 2.1 thus tells us how the distribution of B and S will change under an intervention on B.) The difference between conditioning and intervening thus corresponds to the difference between two questions: What would it be reasonable for me to predict regarding the value of S when I observe the value of B, assuming that no intervention occurs and whatever system has been generating the values of B and S remains intact? What would it be reasonable for me to predict regarding the value of S, if I were to physically manipulate the value of B in the manner described above?

To say that there is an important difference between conditioning and intervening is not of course to say that there are no systematic connections between these two notions. Indeed to a very large extent, *the* problem of causal inference, at least in non-experimental contexts, is when (and how) it is possible to infer from information about conditional probability relationships to claims about what would happen under possible interventions—an issue to which I will turn in Section 6 below. However, from the perspective of a manipulablity theory, the structure of this problem is fundamentally obscured if we do not distinguish between conditioning and intervening.

It should also be clear from the above characterization that interventions function in broadly the same way as Lewisian miracles. When we consider, within Lewis' framework, a counterfactual like (4.4) "If the barometer reading had been low, a storm would have occurred" we imagine that the antecedent of this counterfactual is made true by the insertion of a small, localized miracle which decouples the value of the barometer reading from the value of the atmospheric pressure. This miracle makes B independent of S and this in turn prevents the sort of backtracking reasoning that might be used to argue for the truth of (4.4): (If the reading was low, that must be because A was low, in which case the storm would have occurred). Like an intervention on B, the insertion of such a miracle gives B an independent causal history and (at least in many cases) this is enough to insure that any change in the value of S is due to the value of B and hence that B is a cause of S.[7] Of course real-life interventions need not literally involve miracles, but we may think of this language as a picturesque way of expressing the idea that an intervention involves a change that comes into the system from the outside and disrupts

---

[7] As the phrase in parentheses suggests, while Lewisian miracles often function in the same way as interventions, they do not always do so. See Woodward (forthcoming) for discussion. One important difference between Lewis' theory and the interventionist approach is that the latter assigns an important role to counterfactuals concerning what will happen under combinations of interventions. These have no direct counterpart in Lewis' theory.

endogenous causal relationships. I thus suggest that Lewis' framework works as well as it does because it tracks the same sorts of relationships that are picked out by a manipulationist or intervention-based theory. In other words, Lewis' "unnatural" semantics for counterfactuals makes good scientific sense when motivated by ideas about the connection between causation and manipulation[8].

5 Direct Causes

I said above that directed graphs and systems of equations convey information about direct causal relationships. Given the notion of an intervention, the notion of a direct cause can be understood in manipulationist terms as follows:

> (**DC**) A necessary and sufficient condition for X to be a direct cause of Y with respect to some variable set Z is that there be a possible intervention on X that will change Y (or the probability distribution of Y) when all other variables in Z besides X and Y are also held fixed at some value by interventions.

Intervening to hold the variables in Z fixed at some values while changing X by means of an intervention means that the variables in Z are set to those values by a process that satisfies the conditions for an intervention while X is changed by some other process, also satisfying the conditions for an intervention, that is causally independent of and uncorrelated with the process that changes X.

As an illustration consider the following pairs of equations and corresponding graphical structures (error terms have been suppressed for expository convenience).

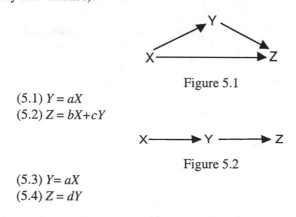

Figure 5.1

(5.1) $Y = aX$
(5.2) $Z = bX + cY$

Figure 5.2

(5.3) $Y = aX$
(5.4) $Z = dY$

---

[8] For similar observations, see Pearl (2000).

According to both (5.1-5.2) and (5.3-5.4), an intervention on X will change Y, an intervention on X will change Z and an intervention on Y will change Z. Nonetheless, (5.1-5.2) and (5.3-5.4) are associated with different graphical structures and make different claims about direct causal relationships. According to (5.1-5.2) X affects Z by two different routes, a direct route and an indirect route that goes through Y. By contrast, according to (5.3-5.4), X affects Z only by a single route that goes through Y. (5.1-5.2) claims that X is a direct cause of Z while (5.3-5.4) denies this. The definition **DC** captures this difference. According to (5.1-5.2), if we intervene to fix the value of Y and then intervene to change the value of X, the value of Z will change—hence X is a direct cause of Z[9]. By contrast, if (5.3-5.4) is correct, then, if we fix the value of Y (at any value), no intervention on X will change Y—hence X is not a direct cause of Y. In other words, while (5.1-5.2) and (5.3-5.4) agree about what will happen to Z under single interventions on either X or Y, they differ in what they predict about what will happen under *combinations* of interventions. In particular, we could determine whether (5.1-5.2) or (5.3-5.4) is the correct structure by doing an experiment in which the value of X is changed and the value of Z observed while the value of Y is held fixed.

DC requires that "all other variables in Z besides X and Y are held fixed at *some* value by interventions". This formulation is needed because some causal relationships are non-linear. Since the relationships in (5.1-5.2) and (5.3-5.4) are linear, the effect of a change in the value of X on Y when Z is fixed will be the same, regardless of the value at which Z is fixed. When relationships are non-linear, this will not be the case. Suppose that F is a variable that takes the values 0 or 1, depending on whether a fire occurs, S is a variable taking the values 0 or 1 depending on whether a short circuit occurs, and O is a variable that takes the values 0 or 1 depending on whether oxygen is present. Suppose that the causal relationship between these variables may be represented by means of the equation

(5.5) $F = S \cdot O$

That is, a fire will occur when and only when both the short circuit and oxygen are present. If $O = 0$, an intervention that changes the value of S will not change the value of F. The formulation of **DC** nonetheless allows S to qualify as a direct cause of F; S is a direct cause of F because there is some value of O (namely $O = 1$) such that with O fixed at that value, a change in S will change F.

Plainly not all causes are direct causes. What is the connection between causation and direct causation? Consider the following candidates for neces-

---

[9] Obviously, this reasoning assumes that the system (5.1-5.2) is modular and in particular that fixing the value of Y in (5.1) does not change (5.2).

sary and sufficient conditions for X to be a cause of Y, which I label (**SC**) and (**NC**) respectively.

> (**SC**) If there is a possible intervention that changes the value of X such that carrying out this intervention will change the value of Y, or the probability distribution of Y, then X causes Y.

> (**NC**) If there are no possible interventions that can change the value of X, or if for all possible interventions that change the value of X, the value of Y (or the probability distribution of Y) does not change, then X does not cause Y.

I believe that (**SC**) is extremely plausible; indeed one may take it to be one of the core commitments of a manipulability theory of causation. By contrast, on the supposition that direct causes are causes, **NC** conflicts with **DC**. To see this consider, a structure like (5.1-5.2) with $b = -ac$. In this structure, the influence of X on Z along the direct and indirect routes will "cancel". In such a structure X is a direct cause of Y according to **DC** and hence (we are supposing) a cause. Nonetheless, there are no interventions on X that will change Y and hence according to **NC**, X fails to cause Y.

This example shows that we need to distinguish between two notions of "cause"[10]. Let us say that X is a *total cause* of Y if and only if it has a non-null total effect on Y—that is, if and only if there is some intervention on X alone (and no other variables) such that for some value of those other variables, this intervention on X will change Y. The *total effect* of a change dx in X on Y is then the change in the value of Y that would result from an intervention on X alone that changes it by amount dx (given the values of other variables that are not descendants of X.) For example, in (5.1-5.2) the total effect on Z of a change of dx in X is $(b+ac)\ dx$ and the total effect on Z of a change dy in Y is cdy. Let us say that X is a *contributing cause* of Y if and only if it makes a non-null contribution to Y along some directed path in the sense that there is some set of values of variables that are *not* on this path such that if these variables were fixed at those values, there is some intervention on X that will change the value of Y.[11] The *contribution* to a change in the value of Y due to

---

[10] For a similar distinction, see Hitchcock (forthcoming).

[11] Why not simply say that X is a contributing cause of Y if and only if X is an ancestor of Y where the ancestor relationship is defined in terms of **DC**? The difficulty with this suggestion is that even if X is an ancestor of Y, it is possible that there are no interventions on X that will change the value of Y, for any values of variables that are not on the directed path from X to Y. For example, suppose that Z is an intermediate variable on the path from X to Y, and that the functions linking X to Z and Z to Y compose in such a way that some intervention on X will change Z for some values of off path variables (so that X is a contributing cause of Z) and some intervention on Z will change Y (so that Z is a contributing cause of Y) but that none of the changes in Z that might be produced by changes in X are such that they will produce changes in Y. (Many of the counterexamples to the transitivity of causation in the philosophi-

a change dx in the value of X (or the effect on Y contributed by this change in X) along some directed path is the change in the value of Y that would result from this change in X, given that the values of off path variables are fixed by independent interventions. For example, if we were to add a third equation (5.6) to (5.1-5.2) relating an additional variable W to X ((5.6)$X = eW$)—that is, if we were to draw an additional arrow from W into X—then although W is not a direct cause of Z, W will be a contributing cause of Z since, freezing the value of Y, there are interventions on W that will change Z. In particular, the contribution (along the path $W \dashrightarrow X \dashrightarrow Z$) to the value of X due to changing the value of W by amount dW is just $ebdw$. When relationships are linear, as in the above examples, it does not matter, for the purposes of identifying either the total effect on Y of a change in the value of X or the contribution this change makes to Y along some route, what values "other" variables assume. When relationships are nonlinear, both total and contributed effect will be relative to the values of other variables. For example, with $O = 0$, a change in the value of S will have no total (or contributed) effect on F. With $O = 1$, the total (and contributed) effect of a change in S from 0 to 1 will be to change F from 0 to 1.

In the case of (5.1-5.2) with $b = -ac$, X is not a total cause of Y but it is a contributing cause. Total causes will satisfy both **SC** and **NC**; contributing causes will satisfy **SC** but need not satisfy **NC**. Direct causes are always contributing causes but contributing causes need not be direct. For example, when the third equation (5.6) is added to (5.1-5.2), W is a contributing although not a direct cause of Z. Both directed graphs and equations aim, in the first instance, at the representation of direct rather than total causal relationships. If we have full information about the functional relationships that represent direct causal relationships, we may recover total causal relationships from this, as illustrated in some of the examples above, but directed graphs, by themselves do not convey such information.

I have argued elsewhere (Woodward forthcoming) that there is an important sense in which the notion of a direct cause (and more generally the notion of a contributing cause) is more fundamental than the notion of a total cause but in what follows I will assume only that the notion of a direct cause is *a* (not necessarily the only) legitimate notion of cause. This assumption is, I believe, implicit in the use of directed graphs and systems of equations to represent causal relationships and also follows from the underlying logic of a manipulability approach to causation. Even when $b = -ac$ in (5.1-5.2), one may still use X to change or manipulate Z—all that one has to do is to fix Y at

---

cal literature have this sort of structure.) In this sort of case, X is not a means for changing Y and it follows from the general connection between causation and manipulation assumed in the manipulablity theory that X is not a contributing (or total) cause of Y.

some value and then wiggle X. Thus there is a perfectly good sense in which X remains a means to changing Z and this, I claim, is enough to establish that there is a legitimate sense in which X is a cause of Z. That sense is captured by the notion of a direct (or contributing) cause.

6 Direct Causes and Probabilities

As explained above, one reason why we need the notion of a direct cause is to capture or represent facts about what will happen under combinations of interventions. Such facts are not always captured by information about total causal relationships, when these are understood as satisfying **SC** and **NC**. For example, in both (5.1-5.2) and (5.3-5.4) X is a total cause of Y, Y is a total cause of Z and X is a total cause of Z. Nonetheless X, Y and Z differ in the direct causal relationships in which they stand in (5.1-5.2) and (5.3-5.4). In this section I will argue that we also need the notion of a direct cause for another reason: in order to formulate plausible conditions connecting causal claims to claims about conditional probability.

I will focus on one of the best known proposals about this connection, the condition (**CC**) formulated by Nancy Cartwright (1983, p. 26) (Broadly similar proposals are endorsed by a number of other writers including Eells (1991) and Eells and Sober (1983). According to (**CC**)

   C causes E iff $\Pr(E/C.K_j) > Pr(E/K_j)$ for all state descriptions K_j over the set $\{C_i\}$ where $\{C_i\}$ satisfies

   (i) If C_i is in $\{C_i\}$, then C_i causes either E or not E.
   (ii) C is not in $\{C_i\}$
   (iii) For all D, if D causes E or D causes not E, then either $D = C$ or D is in $\{C_i\}$
   (iv) If C_i is in $\{C_i\}$, then C does not cause C_i.

CC can be interpreted in at least two ways—as a condition on total causes and as a condition on contributing causes. In what follows I will assume the latter interpretation (i. e., I will use "cause" to mean "contributing cause"), unless explicitly indicated otherwise. **CC** differs from the characterizations of causation considered above in a number of respects. First, **CC** requires that causes (or better, a change in the value of the cause variable from absent to present) raise the probabilities of their effect. By contrast, both the contributing and total notions of cause described above require only that a change in the value of the cause variable change the value or the probability distribution of the effect variable. This difference strikes me as largely (although perhaps not entirely) terminological. **CC** attempts to capture the notion of a positive causal factor or of a promoting cause as opposed to the notion of a negative causal factor or a preventive or inhibiting cause. By contrast, **DC**, as well as **NC** and **SC,** attempt to capture the broader notion of one variable's being causally relevant (either positively or negatively) to another. A variety of considerations of convenience seem to me favor this broader usage[12].

Second, **CC** imposes what has come to be called a unanimity requirement: C causes E if and only if C raises the probability of E across *all* background contexts K (or all situations that are "otherwise causally homogenous with respect to E" (Cartwright 1983, p. 25)). Several commentators (e.g., Dupre 1984) have objected that this requirement is too strong on the grounds that it has unintuitive consequences. For example, it requires that we withdraw the claim that "smoking causes lung cancer among human beings" if we were to discover even a small subpopulation in which, perhaps as a result of some genetic quirk, smoking fails to raise the probability of lung cancer. I agree with this objection but think that we do not need to leave matters at the level of intuition. If I am correct in claiming that the underlying point or rationale of our classifying the relationship between X and Y as causal or noncausal has to do with whether or not X is a potential means for controlling or manipulating Y, then there is little motivation for the unanimity requirement, since even if we agree to restrict the notion of cause to mean "positive or promoting cause", it is clear that X can be used to manipulate Y in a way that is positive for Y even if the unanimity requirement is violated. Instead what the manipulability conception (and in particular, **SC**) suggests is something like the following: X will be means for manipulating Y in positive way (i.e., for promoting Y), if as there is at least one background context in which X raises the probability of Y.

If we look for a connection between causation and facts about probabilities that is in the spirit of **CC** but incorporates these two points, it will be a proposal of the following form: X causes Y if and only if X and Y are dependent conditional on certain other factors F. The problem of finding a connection between causation and probability then becomes one of specifying what these other factors F are. In other words, the question is this: what should be held fixed (that is, conditioned on) if the conditional dependence of C on E is to be used as a test for whether C causes E? **CC** says that the other factors F that should be conditioned on are all other causes of E with the exception of

---

[12] A minor terminological annoyance is that to assess whether X causes Y, other factors that are negatively as well as positively causally relevant to Y must be controlled for. Thus if "cause" is restricted to mean "positive cause", it is incorrect to say that the only factors that need to be controlled for are "other causes of Y". One needs some additional vocabulary to describe the other factors that need to be controlled for. A more fundamental difficulty is that once one moves away from cause variables that are binary valued, it often becomes unclear what value of the cause variable corresponds to the "absence" of the cause and hence what the state is in comparison with which the "presence" of the cause should raise the probability of the effect.

those causes of E that are on the causal chain from C to E. (As Cartwright explains (1983, p. 26) condition (iv) in **CC** is intended to exclude conditioning on such factors). The motivation for not holding fixed causal factors that are between C and E may seem obvious. If we are dealing with a causal structure like that represented in (5.3-5.4) and Figure 5.2 in which there is a single directed path from X to Z with Y as a causally intermediate variable, then one would expect that conditional on Y, X and Z will be independent. Hence, if Y is one of the background factors on which we condition when we test for whether X causes Z we will reach the mistaken conclusion that X does not cause Z. However, as Cartwright herself recognizes (1983, p.30, 1989, pp. 95ff), the claim that, as (iv) requires, we should *never* control for such intermediate variables is too strong[13]. Suppose that we are presented with a triangular structure like that in (5.1-5.2) and Figure 5.1 in which both X and Y are direct causes of Z and X is also a direct cause of Y. Clearly if the direct causal connection between X and Z is to reveal itself in the probabilistic dependence of Z on X conditional on some appropriately chosen set of other factors, these other factors must include Y which is causally intermediate between X and Y. That is, to capture the direct influence of X on Z, we must in some way control for or correct for the influence of Y on Z. Moreover, since as we have seen the total cause structure is the same in both (5.1-5.2) and (5.3-5.4), we need to know the direct causal relationships (and not just the total causal relationships) between X, Y and Z in order to know what to control for when we test for, e. g., whether X is a cause of Z.

What these examples bring out is that in determining what should be controlled for or conditioned on for the purposes of assessing whether X causes Z, we need more than information about the other causes (in either the contributing or total sense) of Z besides X. We also need to know how, as it were, those other causes are connected up—with one another, with X and with Z. It is just this information about direct causal relationships that is contained in the associated equational or directed graph structure and this in turn suggests that information about such structures is essential if the sort of project represented by **CC** (the project of formulating systematic relationships between causal claims and conditional probability relationships) is to have any hope of success.

This point of view is also reflected in the so-called causal Markov condition (**CM**). This is a generalization of familiar ideas about screening off,

---

[13] Cartwright (1989, p. 96) abandons requirement (iv); replacing it with a requirement involving information about singular causal processes. I fully agree with Cartwright that some additional information is needed to distinguish what needs to be controlled for in structures like (5.1-5.2) and (5.3-5.4). However, my suggestion is that what is needed is rather additional information about direct causal relationships at the type-level.

first formulated by Reichenbach (1956), and has figured heavily in recent work on causal inference by Judea Pearl (2000) and by Clark Glymour and his associates (Spirtes, Glymour and Scheines 1993). **CM** says that conditional on its parents or direct causes, every variable is independent of every other variable except its effects:

(**CM**) For all Y distinct from X, if X does not cause Y, then $Pr(X/Parents\,(X)) = Pr(X/Parents(X)\,.Y)$

As with **CC**, the word "cause" in **CM**, can be interpreted as either "contributing cause" or "total cause". ("Parents" must of course be understood as meaning "direct cause"). I will assume the former interpretation unless indicated otherwise, although I think that insofar as **CM** is plausible at all, it is equally plausible under either interpretation. Like **CC**, **CM** connects claims about causal relationships to facts about relationships between conditional probabilities. However, unlike **CC**, **CM** is formulated in terms of information about the direct causes of X. As we will see, this is a crucial difference.

It is well known that there are circumstances under which **CM** fails to hold. For example **CM** will be violated if purely accidental correlations that reflect no causal connections at all occur. **CM** can also break down in the presence of cyclic causal relationships or when variables are measured imperfectly or when their values are drawn from mixtures of distinct probability distributions[14]. Although the point is not widely appreciated, **CC** will also be violated in all of these circumstances. However, when such circumstances are excluded, there is a plausible case to be made that **CM** follows from a manipulability conception of causation—or so Dan Hausman and I have argued elsewhere (Hausman and Woodward 1999). **CM** thus has some claim to be regarded as a conceptual truth about causation. Moreover, although I will not attempt to argue for this claim here, I conjecture that insofar as there is *any* systematic connection between causation and conditional independence relationships in acyclic causal systems, it is captured by **CM**. That is, to the extent that **CM** fails to hold, *no* general test for causation—neither **CC** nor any competitor—formulated in terms of conditional independence relationships will work. This gives us a reason for focusing on **CM** and asking what if anything it implies about the connection between casual claims and conditional probabilities.

Contraposing **CM** gives a *sufficient* condition for causation in terms of conditional dependence relationships: assuming that Y is distinct from X, if $Pr(X/Parents\,(X)) \neq Pr\,(X/Parents\,(X).Y)$ then X causes (i. e., is a contributing cause of) Y. However, the converse of this claim is *not* correct, if "cause"

---

[14] For a discussion of the circumstances in which **CM** fails, see Hausman and Woodward (1999).

means "contributing cause". In (5.1-5.2) and Figure 5.1, with $b = -ac$, X and Z are uncorrelated and hence $Pr(X/Parents(X)) = Pr(X/Parents\ (X).\ Z)$. Nonetheless X is a contributing cause of Z. We would have a necessary condition for contributing causation if we were willing to assume the converse of **CM**, which Spirtes, Glymour and Scheines (1993) call Faithfulness (**F**).

> (**F**): If X causes (is a contributing cause of) Y, then $Pr\ (X/Parents\ (X) \neq Pr\ (X/Parents\ (X).Y)$

Taken together, **CM** and **F** provide necessary and sufficient conditions, formulated in terms of facts about relationships between conditional probabilities for X to be a contributing cause of Y. Interestingly, the conjunction **CM.F** is a condition that is rather different in form from **CC**, even allowing for the fact that the notion of direct causation plays no role in **CC**. While **CC** looks (roughly) at whether X and Y are dependent conditional on all *other* causes of Y (besides X) that meet certain additional conditions, **CM.F** asks whether X and Y are dependent conditional on *all direct* causes of X.

What justification is there for assuming (**F**)? Spirtes, Glymour and Scheines (1993) advance a measure-theoretic argument: given certain assumptions, violations of Faithfulness will be rare in the measure-theoretic sense. However, this is at best a reason for assuming faithfulness in causal inference problems—assuming faithfulness will only rarely mislead us. As Spirtes, Glymour and Scheines readily concede, "rare" does not mean impossible. To the extent that our interest is in giving a condition that is strictly necessary for it to be true that X causes Y (and not just a fallible test for whether X causes Y), assuming **F** is problematic. In contrast to **CM**, even those who advocate **F** do not suppose that it has any claim to be regarded as a conceptual truth about causation.

Is there some other candidate for a necessary condition that is more plausible than **F**? Although I lack the space for a detailed exploration of the possibilities, I think that there are reasons for skepticism. For example, the suggestion that a necessary condition for X to be a contributing cause of Y is that Y be dependent on X conditional on all direct causes of Y that are not identical with or descendants of X fails for several reasons, including the possibility of failures of faithfulness. The fundamental problem is that **CM** says merely that if certain causal relationships hold, then certain conditional independence relationships follow. **CM** doesn't say that if these causal relationships hold, then *only* these conditional independence relationships and no others hold[15]. However, something like this latter assumption seems required

---

[15] As (5.1-5.2) show, it is perfectly possible for independence relationships that do not follow from **CM** to hold because of, e. g., cancellations among coefficients in equations. It is cases of this sort that constitute violations of faithfulness.

if we are to have a necessary condition connecting causation and conditional probabilities. For such a necessary condition to hold, it would need to be true not just that (a) some inequality between conditional probabilities always indicates the presence of a causal relationship but that (b) causal relationships always reveal themselves in some inequality between conditional probabilities. While there is some reason to think that (a) is built into our concept of causation, it is hard to see how such an argument could be made on behalf of (b).

Does this assessment change if instead we look for necessary and sufficient conditions for X to be a total cause (rather than a contributing cause) of Y that are formulated in terms of conditional dependence relationships? As suggested above, **CM** yields a plausible sufficient condition for X to be a total cause of Y: (**TSC**) If Y is distinct from X and $Pr(X/Parents\ (X)) \neq Pr(X/Parents\ (X)).\ Y)$ then X is a total cause of Y. However, we cannot replace the reference to $Parents(X)$ in **TSC** with some condition formulated in terms of total causes—that is, the condition will fail to be sufficient if we fail to control for direct causes of X that are not total causes. As an illustration, consider the following structure

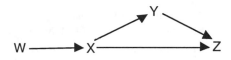

(5.1) $Y = aX$
(5.2) $Z = bX+cY$
(6.1) $W = dX$

Suppose as before that $ac= -b$ Then X is not a total cause of Z. Instead, Y is the only total cause of Z. Assuming **CM**, W is independent of Z conditional on both X and Y, but W is *not* independent of Z conditional just on Y. Nonetheless W is not a total cause of Z.[16] The inference from the fact that Z and W are dependent conditional on all of the total causes of Z to the conclusion that W causes Z is mistaken.

Formulating a sufficient condition for total causation in terms of conditional dependence relations thus requires information about direct causal relationships and this is an additional reason for thinking that the notion of a direct cause is an indispensable one. Moreover, again because of the possibility of violations of faithfulness, the converse of **TSC** does not hold. Consider a structure in which X and Y are the only direct causes of Z, Y is exogenous, and X is not, and Y is not a cause of X Then with right values of the coefficients linking X to Z and Y to Z, it is possible for X to be independent of Z

---

16 Thanks to Chris Hitchcock for supplying this example.

and hence independent of Z conditional on *Parents (X)*. Nonetheless, in this structure X is a total cause of Z.

In summary, we may draw two more general conclusions. First, we require the notion of direct causation if we are to formulate any plausible connection between causation (whether contributing or total) and probability. Second any defensible connection between causation and probability is likely to involve only a sufficient rather than a necessary and sufficient condition. In this sense, the connection will be far weaker than the necessary and sufficient conditions sought in the philosophical literature on probabilistic causation. If we want necessary and sufficient conditions for causation that apply even in circumstances in which **CM** is violated, a counterfactual approach is more promising.

References

Cartwright, N. (1983) *How the Laws of Physics Lie*. Oxford: Oxford University Press.

Cartwright, N. (1989) *Nature's Capacities and Their Measurement*. Oxford: Clarendon Press.

Cook, T. and Campbell, D. (1979) *Quasi-Experimentation: Design and Analysis Issues for Field Settings*. Boston: Houghton Miflin Company.

Eells, E. (1991) *Probabilistic Causality*. Cambridge: Cambridge University Press.

Eells, E. and Sober, E. (1983) Probabilistic causality and the question of transitivity. *Philosophy of Science,* 50, 35-57.

Hausman, D. (1998) *Causal Asymmetries*. Cambridge: Cambridge University Press.

Hausman, D. and Woodward, J. (1999) Independence, invariance, and the causal Markov condition. *British Journal for the Philosophy of Science,* 50, 521-583.

Hitchcock, C. (forthcoming) *The Intransitivity of Causation Revealed in Equations and Graphs*.

Lewis, D. (1973) Causation. *Journal of Philosophy* 70, 556-567. Page references in text to reprinting in Lewis (1986).

Lewis, D. (1979) Counterfactuals dependence and time's arrow. *Nous* 13, 455- 76. Page references in text to reprinting in Lewis (1986).

Lewis, D. (1986) *Philosophical Papers,* Vol. 2, New York: Oxford University Press. Vol. 63, 113-37.

Meek, C. and Glymour, C. (1994) Conditioning and intervening. *British Journal for the Philosophy of Science* 45, 1001-1021.

Pearl, J. (1995) Causal diagrams for empirical research. *Biometrika* 82, 669-688.

Pearl, J. (2000) *Causation: Models, Reasoning and Inference*. Cambridge: Cambridge University Press.

Price, H. (1991) Agency and probabalistic causality. *British Journal for the Philosophy of Science* 42 157 -76.

Reichenbach, H. (1956) *The Direction of Time*. Berkeley: University of California Press.

Spirtes, P., Glymour, C. and Scheines, R. (1993) *Causation, Prediction and Search*. New York: Springer-Verlag.

Woodward, J. (1997) Explanation, invariance and intervention. *PSA 1996*, vol 2, S26- 41.

Woodward, J. (1999) Causal interpretation in systems of equations. *Synthese,* 121, 199-247.

Woodward, J. (2000) Explanation and invariance in the special sciences. *British Journal for the Philosophy of Science,* 51, 197-254.

Woodward, J. (forthcoming) Causality and manipulation.

Von Wright, G. (1971) *Explanation and Understanding*. Ithaca, New York: Cornell University Press.

4

Modularity: It Can—and Generally Does—Fail*

NANCY CARTWRIGHT

London School of Economics
University of California, San Diego

1 Introduction

This paper pursues themes developed in my recent book *The Dappled World: A Study of the Boundaries of Science.* (Cartwright 1999). The book is a Scotist book—in accord with the viewpoint of Duns Scotus. It extols the particular over the universal, the diverse over the homogeneous and the local over the global. Its central thesis is that in the world that science studies, differences turn out to matter. Correlatively, universal methods and universal theories should be viewed with suspicion. We should look very carefully at their empirical justification before we adopt them.

The topic in this volume is causality; I shall defend a particularist view of our subject. Causal systems differ. What is characteristic of one is not characteristic of all and the methods that work for finding out about one need not work for finding out about another. I shall argue this here for one specific characteristic: *modularity.* Very roughly, a system of causal laws is modular in the sense I shall discuss when each effect in the system has one cause all its own, a cause that causes it but does not cause any other effect in the system. On the face of it this may seem a very special, probably rare,

---

* Research for this paper was supported by a grant from the Latsis Foundation.

Stochastic Causality.
Maria Carla Galavotti, Patrick Suppes and Domenico Costantini (eds.).
Copyright © 2001, CSLI Publications.

situation. But a number of authors currently writing on causality suppose just the opposite. Modularity, they say, is a universal characteristic of causal systems. I shall argue that they are mistaken.

2   What Is Modularity?

Behind the idea that each effect in a causal system[1] should have a cause of its own is another idea, the idea that each effect in the system must be able to take any value in its range consistent with all other effects in the system taking any values in their ranges. There are two standard ways in which people seem to think this can happen; it will be apparent that different senses of *able* are involved.

In the first place, a second collection of causal systems very similar to the first may be possible in which all the laws are exactly the same except for the laws for the particular effect in question. In the new systems these laws are replaced by new laws that dictate that the effect take some specific value, where the systems in the collection cover all the values in the range of that effect.

This interpretation clearly requires that we be able to make sense of the claim that an alternative set of laws is possible. For my own part I have no trouble with this concept: in *The Dappled World* I argue that laws are not fundamental but instead arise as the result of the successful operation of a stable arrangement of features with stable capacities. Nevertheless, I do not see any grounds for the assumption that the right kind of alternative arrangements must be possible to give rise to just the right sets of laws to make modularity true. At any rate this way of securing modularity is not my topic in this paper.

The second way in which modularity might obtain is when each effect in the system has a cause all of its own that can contribute to whatever its other causes are doing to make the effect take any value in its range. This is the one I will discuss here. I will also in this paper restrict my attention to systems of causal laws that are both linear and deterministic. In this case the commitment to modularity of the second kind becomes a commitment to what I call "epistemic convenience".

An *epistemically convenient linear deterministic system* is a system of causal laws of the following form[2]

---

[1] I shall use "causal system" to refer to a set of causal laws and "causal structure" to refer to a set of hypotheses about causal laws.

[2] Somewhat more accurately, I should say "a system of laws generated by laws of the following form", for I take it that causal laws are transitive. For a more precise formulation, see Cartwright (2000).

$$x_1 \ c= u_1$$
$$x_2 \ c= a_{21}x_1 + u_2$$
$$\vdots$$
$$x_n \ c= \Sigma a_{nj}x_j + u_n$$

plus a probability measure $P(u_1,\ldots,u_n)$, where

i.   there are no cross restraints among the u's[3] and the u's are probabilistically independent of each other;

ii. for all j, $Prob(u_j = 0) \neq 1$.

The symbol "c=" shows that the law is a causal law. It implies both that the relation obtained by replacing "c=" with "=" holds and that all the quantities referred to on the right-hand side are causes of the one on the left-hand side.

These systems are epistemically convenient because they make it easy to employ randomized treatment/control experiments to settle questions about causality using the method of concomitant variation. I will explain in more detail below, but the basic idea can be seen by considering the most straightforward version of the method of concomitant variation: to test if x_j causes x_e and with what strength, use u_j to vary x_j while holding fixed all the other u's, and look to see how x_e varies in train. Conditions i) and ii) guarantee that this can be done.

A number of authors from different fields maintain that modularity[4] is a universal characteristic of causal systems. This includes economic methodologist Kevin Hoover[5], possibly Herbert Simon (1977), economists T. F. Cooley and Stephan LeRoy (1985), Judea Pearl (2000) in his new study of counterfactuals, James Woodward (1997), Daniel Hausman (1998) and Daniel Hausman and James Woodward (1999) jointly in a paper on the causal Markov condition. I aim to show that, contrary to their claims, we can have causality without modularity. I focus on the second kind of modularity here in part because that is the kind I have found most explicitly

---

[3] See n. 16.

[4] Or some closely related doctrine. Much of what I say can be reformulated to bear on various different versions of a modularity-like condition.

[5] See his defense of the invariance of the conditional probability of the effect on the cause in Hoover (forthcoming). In this discussion Hoover seems to suppose that there always is some way for a single cause to vary and to do so without any change in the overall set of laws. At other places, I think, he does not assume this. But he does speak with approval of Herbert Simon's way of characterizing causal order, and Simon's charcterization requires the possibility of separate variation of each factor.

defended. Hence I shall be arguing that *not all causal systems are epistemically convenient.*[6]

3   The Method of Concomitant Variation

We *say* that the method of concomitant variation is a good way to test a causal claim. But can we show it? For an epistemically convenient system we can, given certain natural assumptions about causal systems. That is one of the best things about an epistemically convenient system—we can use the method of concomitant variation to find out about it.

I shall not give the proof here, but rather describe some results we can show.

Here are the assumptions I shall make about linear deterministic systems of causal laws:

A_1: *Functional dependence.* Any causal equation re presents a true functional relation.

A_2: *Anti-symmetry and irreflexivity.* If q causes r, r does not cause q and q does not cause q.

A_3: *Uniqueness of coefficients.* No effect has more than one expansion in the same set of causes.

A_4: *Numerical transitivity.* Causally correct equations remain causally correct if we substitute for any effect any function in its causes that is among nature's causal laws.

A_5: *Consistency.* Any two causally correct equations for the same effect can be brought into the same form by substituting for effects in them functions of their causes given in nature's causal laws.

A_6: *Generalized Reichenbach principle.* No quantities are functionally related unless the relation follows from nature's causal laws.

The result I shall describe says very roughly that when the underlying linear deterministic system is an epistemically convenient one, then a causal hypothesis is correct iff the method of concomitant variation says it is so. In order to express this more precisely, we shall have to know what the form of the causal hypotheses in question are, what it is for hypotheses of this

---

6 The authors mentioned here all have slightly different views, formulated and defended differently and with different caveats. I apologize for lumping them all together. Clearly not all the remarks I make are relevant to every view. In fact I will focus on a very specific form of the claim for universal modularity. Nevertheless, most of what I say can be translated to apply to other forms of the claim.

form to be causally correct and what it is to pass the test for concomitant variation.

The usual hypotheses on offer when we suppose the underlying causal system to be linear and deterministic[7] are in the form of regression equations:

R: x_k c= $\Sigma a_{kj} x_j + \Psi_k$, for $\Psi_k \perp x_j$ for all j,
where $x \perp y$ means that $\langle xy \rangle = \langle x \rangle \langle y \rangle$.

What exactly does this hypothesis claim? I take it that the usual understanding is this: every quantity represented by a named variable (an "x") on the right-hand side is a genuine cause of the quantity represented on the left-hand side, and the coefficients are "right". The random variable "Ψ" represents a sum of not-yet-known causes that turn R into a direct representation of one of the laws of the system. So I propose to define correctness thus: an equation of the form R: x_k c= $\Sigma a_{kj} x_j + \Psi_k$ ($1 \leq j \leq m$), for $\Psi_k \perp x_j$, is *correct* iff there exist $\{b_j\}$ (possibly $b_j = 0$), $\{q_j'\}$ such that q_k c= $\Sigma a_{kj} q_j + \Sigma b_j q_j' + u_k (1 \leq j \leq m)$, where q_j does not cause q_j'. (This last restriction ensures that all the omitted factors are causally antecedent to or "simultaneous" with those mentioned in the regression formula. Note, x_j represents q_j.)

Now let us consider *concomitant variation*. In an epistemically convenient linear deterministic system, the value of the x's are fixed by the u's, and the u's can vary independently of each other. The core idea of the method is to take the concomitant variation between x_o and x_e when u_c varies while all the other u's are fixed as a measure of the coefficient of x_c in nature's equation for x_e.

To state the relevant theorems we shall need some notation. Let $\Delta j(\alpha)x_n$ $=df x_n(u_1 = U_1, \ldots , u_{j-1} = U_{j-1}, u_j = U_j + \alpha, u_{j+1} = U_{j+1}, \ldots , u_m = U_m) - x_n(u_1 = U_1, \ldots , u_{j-1} = U_{j-1}, u_j = U_j, u_{j+1} = U_{j+1}, \ldots , u_m = U_m)$. Then we can prove[8]

Theorem 1. A (true) regression equation x_k c= $\Sigma_{j=1}^{k-1} a_{kj} x_j + \Psi_k$, is causally correct iff for all values of α and J, $1 \leq J \leq k$, $\Delta_J(\alpha)x_k = \Sigma a_{kj} \Delta_J(\alpha)x_k$; i.e. iff the equation predicts rightly the differences in x_k generated from variations in any right-hand-side variable.

Notice, however, that this theorem is not very helpful to us in making causal inferences because it will be hard to tell whether an equation has indeed predicted the differences rightly. That is because we will not know

---

[7] Or when we are prepared to model the system as linear and deterministic for some reason or another.

[8] For a proof of the three theorems see Cartwright (2000, Ch. 3, also forthcoming in *Philosophy of Science*.). The formalization and proofs are inspired by the work of James Woodward on invariance, which argues more informally for a more loosely stated claim. I am aiming to make these claims more precise in these theorems.

what $\Delta_j(\alpha)x_j$ should be unless we know how variations in u_j affect x_j and to know that we will have to know the causal relations between x_j and x_j. So in order to judge whether each of the x_j affects x_k in the way hypothesized, we will have already to know how they affect each other. If we happen to know that none of them affect the others at all, we will be in a better situation, since the following can be trivially derived from the previous theorem:

Theorem 2. A (true) regression equation for x_k in which no right-hand side variable causes any other is causally correct iff for all α and J, $\Delta_j(\alpha)x_k = a_{kj}\Delta_j(\alpha)u_j$.

We can also do somewhat better if we have a complete set of hypotheses about the right-hand-side variables. To explain this, let me define a *complete causal structure* that represents an epistemically convenient linear deterministic system with probability measure, as a triple, $<X = \{x_1,\ldots,x_n\}$, μ, CLH>, where μ is a probability measure over X and where the causal law hypotheses, CLH, have the following form:

$$x_1 c = \Psi_1$$
$$x_2 c = a_{21}x_1 + \Psi_2$$
$$\vdots$$
$$x_n c = \sum_{j=1}^{n-1} a_{nj}x_j + \Psi_n,$$

with $\Psi_j \perp x_k$, for all $k < j$. In general $n < m$, where m is the number of effects in the causal system. Now I can formulate

Theorem 3. If for all x_k in a complete causal structure, the $\Delta_j(\alpha)x_k$ that actually obtains equals $\Delta_j(\alpha)x_k$ *as predicted by the causal structure* for all α and J, $1 \leq J \leq n$, then all the hypotheses of the structure are correct.

I take it that it is the kind of facts recorded in these theorems that make epistemically convenient systems so desirable, so that we might wish—if we could have it—for all causal systems to be epistemically convenient. But is it sensible to think they are? In the next section I will give some obvious starting reasons for thinking the answer must be "no".

4 Three Peculiarities of Epistemic Convenience

To notice how odd the requirement of epistemic convenience is, let us look first at some ordinary object whose operation would naturally be modeled at most points by a system of deterministic laws—for instance a well-made toaster like the one in Figure 1[9]. The expansion of the sensor due to the heat

---

[9] Figure 1 is drawn by Emily Cartwright following an explanation and illustration of the functioning of a toaster in Macaulay (1988).

The Toaster

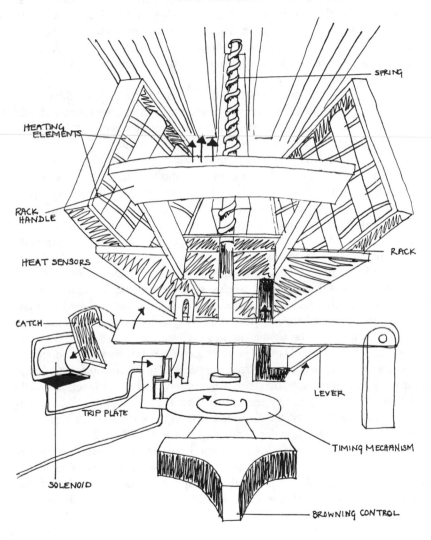

SPRING

HEATING ELEMENTS

RACK HANDLE

HEAT SENSORS

CATCH

TRIP PLATE

SOLENOID

RACK

LEVER

TIMING MECHANISM

BROWNING CONTROL

Figure 1.

produces a contact between the trip plate and the sensor. This completes the circuit, allowing the solenoid to attract the catch, which releases the lever. The lever moves forward and pushes the toast rack open.

I would say that the movement of the lever causes the movement of the rack. It also causes a break in the circuit. Where then is the special cause that affects only the movement of the rack? Indeed, where is there space for it? The rack is bolted to the lever. The rack must move exactly as the lever dictates. So long as the toaster stays intact and operates as it is supposed to, the movement of the rack must be fixed by the movement of the lever to which it is bolted.

Perhaps, though, we should take the bolting of the lever to the rack as an additional cause of the movement of the rack? In my opinion we should not. To do so is to mix up causes that produce effects within the properly operating toaster with the facts responsible for the toaster operating in the way it does; that is, to confuse the causal laws at work with the reason those are the causal laws at work.[10] But even if we did add the bolting together at this point as a cause, I do not see how it could satisfy conditions i) and ii). It does after all happen as part of the execution of the overall design of the toaster, and hence it is highly correlated with all the other similar causes that we should add if we add this one, such as the locating of the trip plate and the locating of the sensor.

The second thing that is odd about the demand for modularity is where it locates the causal nexus. It is usual to suppose that the fact that C causes E depends on some relations between C and E.[11] Modularity makes it depend on the relation between the causes of C and C: C cannot cause anything unless it itself is brought about in a very special way.

Indeed, I think that Daniel Hausman embraces this view: "…people … believe that causes make their effects happen and not vice versa. This belief is an exaggerated metaphysical pun, which derives from the fact that people can make things happen by their causes. This belief presupposes the possibility of intervention and the claim that not all the causes of a given event are nomically connected to one another" (Hausman 1998, 272).

This is a very strong view that should be contrasted with the weaker view, closer (on my reading) to that of Hume, that the concept of causation *arises* because "people can make things happen by their causes", but that this condition does not constitute a truth condition for causation. The

---

[10] For a more complete discussion of this point, see the distinction between nomological machines, on the one hand, and the laws that such machines give rise to, on the other, in Cartwright (1999).

[11] Or perhaps, since C and E here pick out types and not particular events, "between C-type events and E-type events".

weaker view requires at most that sometimes a cause of a cause of a given effect vary independently of all the "other" causes of that effect; it does not take epistemic convenience as universal.[12] In my opinion the weaker view only makes sense as an empirical or historical claim about how we do in fact form our concepts, and about that, we still do not have a reliable account. The stronger view just seems odd.

Thirdly, the doctrine seems to imply that it is impossible to build a bomb that cannot be defused. Nor can we make a deterministic device of this sort: the correct functioning of the mechanisms requires that they operate in a vacuum; so we seal the whole device in a vacuum in such a way that we cannot penetrate the cover to affect one cause in the chain without affecting all of them. Maybe we cannot build a device of this sort—but why not? It does not seem like the claim that we cannot build a perpetual motion machine. On the doctrine of universal epistemic convenience we either have to say that these devices are indeed impossible, or that what is going on from one step to the next inside the cover is not causation, no matter how much it looks like other known cases of causation or passes other tests for causation (such as the transfer of energy/momentum test or the demand that a cause increase the probability of its effects holding fixed a full set of other causes).

Given that the claim to empirical convenience as a universal condition on causality has these odd features, what might motivate one to adopt it? Three motivations are ready to hand: we might be moved by operationalist intuitions, or by pragmatist intuitions or we might be very optimistic about how nicely the world is arranged for us. I will take up each in turn.

5 Motivations for Epistemic Convenience: 'Excessive' Operationalism

This is a hypothesis of Arthur Fine's[13]: Advocates of modularity conflate the truth conditions for a causal claim with conditions which *were they to*

---

[12] It is also surprising that Hausman focuses on the supposition of (something like) epistemic convenience as a necessary condition, but does not stress the equally problematic matter of the possibility of choice. We all know the classic debate about free will and determinism: it looks as if people cannot make things happen by their causes unless the causes of the causes are not themselves determined by factors outside our will, and that in turn looks to preclude universal determinism. If that should follow, it would not trouble me, but many advocates of modularity also defend the causal Markov condition—which I attack—on the grounds of universal determinism. Moreover, the need for *us* to cause some of the causes at least some of the time seems equally necessary whether one takes the strong view that Hausman maintains or the weaker view—which does not require epistemic convenience as a truth condition—that I described as closer to Hume.

[13] Conversation, May, 2000, Athens, Ohio.

obtain would make for a ready test. As we have seen, a central feature of deterministic systems that are epistemically convenient is that we can use the simplest version of the method of concomitant variation within them: to test "x_c cause x_e", consider situations in which x_c varies *without variation in any 'other' causes of x_e* and look for variation in x_e. I think this is particularly plausible as a motivation for economists. Economists in general tend to be very loyal to empiricism, even to the point of adopting operationalism. For instance, they do not like to admit preferences as a psychological category but prefer to use only preferences that are revealed in actions.

In general, versions of operationalism that elevate a good test to a truth condition are in disfavour. Still we need not dispute the matter here, for, even were we disposed to this kind of operationalism in the special case at hand, it would not do the job. Simple concomitant variation is no better test than many others—including more complicated methods of concomitant variation. So operationalism will not lead us to limit causal concepts to systems that admit tests by simple concomitant variation at the cost of other kinds of systems. In particular, the simple method does not demand any 'less'[14] background knowledge than tests using more complicated versions of concomitant variation, which can be performed on other kinds of deterministic systems, nor knowledge of a different kind.

Let me illustrate. We will continue to look at linear deterministic systems and we will still assume that all exogenous factors are mutually unconstrained: there are no functional relations among them.[15] And we will still test for causal relations by the method of concomitant variation.

Imagine then that we wish to learn the overall strength, if any, of x_1's capacity to affect x_e, where we assume we know some cause u_1 that has a known effect (of, say, size b_1) on x_1 and whose variation we can observe. In the general case where we do not presuppose epistemic convenience, every candidate for u_1 may well affect x_e by other intermediaries, say $x_2,...,x_n$, as well. Suppose the overall strength of its capacity to affect x_j is b_j and of x_j to affect x_e is c_j.

We aim to compare two different situations, which are identified by the values assigned to the u's: $S = <U_1,U_2,...,U_m>$ and $S' = <U_1'\ U_1,U_2,...,U_m>$, where the u's constitute a complete set of mutually unconstrained exogenous factors that determine x_e. Then

---

[14] I put "less" in scare quotes because I do not mean to get us involved in any formal notions of more and less information.

[15] If they are constrained, re-express all of them as appropriate functions of a further set of mutually unconstrained factors. Notice that this has no implications one way or another about whether, for example, two endogenous factors share all the same exogenous causes. The point of this is to allow that the exogenous factors *can* vary independently of each other.

$$x_e' - x_e = \Sigma_{j=1}{}^m b_j c_j (U_1' - U_1)$$

or, letting $(\Sigma_{j=2}{}^m b_j c_j)/b_1 =\mathrm{df}\ A$,

$$(x_e' - x_e)/b_1 (U_1' - U_1) = c_1 + A.$$

Now here is the argument we might be tempted to give in favour of epistemically convenient systems. If we have an epistemically convenient system, $A = 0$, so $c_1 = x_e' - x_e/b_1(U_1' - U_1)$. Otherwise we need to know the value of A. So we need less antecedent knowledge if our systems are epistemically convenient.

But clearly the last two sentences are mistaken: $A = 0$ is just as much a value of A as any other; to apply the method of concomitant variation, we need to know (or be willing to bet on) the value of A in any case. *Sometimes* there may be some factor u_1 for which it is fairly easy to know that its effect on x_e by routes other than x_1 is zero. This for example, is, in my opinion, the case with J.L. Mackies's famous hypothesis that the sounding of the end-of-workday hooters in Manchester brings the workers out onto the streets in London. Here we know various ways to make the hooters in Manchester sound of which we can be fairly confident that they could not get the workers in London out except via making the Manchester hooters hoot.[16]

But equally, sometimes we may know for some exogenous factors that do affect x_e by routes other than x_1 what the overall strength of that effect is—if, for instance, we have data on variations in x_e given variations in u_1 when the route from u_1 to x_1 is blocked.

Let us review some of the prominent facts we would need to know for a brute force application of the method of concomitant variation, as I have described it, in a linear deterministic system. To test "x_1 causes x_e with strength c" we need to know

1. of a factor u_1 that it is exogenous to the system under study, that it causes x_1 and with what strength it does so;

2. of a set of factors that they are exogenous, that they are mutually unconstrained, and that together, possibly including u_1, they are sufficient to fix x_e but not sufficient to fix u_1;

3. what would happen to x_e in two different situations for which the values of the exogenous factors described in 2. do not vary, except for the value of u_1, which does vary;

4. the overall strength of u_1's capacity to affect x_e by other routes than by causing x_1.

---

[16] For a discussion, see Cartwright (1989).

My point here is that we need to know (or find a way around knowing) all of this information whether or not the system is epistemically convenient.

Why then have I called these special kinds of systems "epistemically convenient" for use of the method of concomitant variation if we need to know (or find our way around knowing) "the same amount" of information to use the method whether the system is epistemically convenient or not? Because when the system is epistemically convenient, it is a lot easier to use randomized treatment/control experiments. That is why I have called these systems "epistemically convenient"; and it is one of the chief arguments James Woodward (2000, p. 10) gives in favour of the claim that causal systems should be epistemically convenient:

> A manipulationist approach to causation explains the role of experimentation in causal inference in a very simple and straight-forward way: Experimentation is relevant to establishing causal claims because those claims consist in, or have immediate implications concerning, claims about what would happen to effects under appropriate manipulations of their putative causes. In other words, the connection between causation and experimentation is built into the very content of causal claims.

Randomized treatment/control experiments provide us with a powerful tool to find our way around knowing large chunks of information we otherwise would need to know. For the point at issue in this paper, we need to be clear about which features of the stock experimental structure help with which aspects of our ignorance.

Randomization allows us to finesse our lack of knowledge of the kinds of facts described in 2. above. When we are considering the effect of u_1 on x_e, we generally do not know a set of "other" exogenous factors sufficient to fix x_e. But a successful randomization ensures that they will be equally distributed in both the treatment and the control groups. Hence there will be no background correlations between these other factors that might confound our results. Observing the outcome in the two groups allows us to find out (roughly[17]) the information we look for in 3.: what happens under variation in u_1.

But notice that randomization and observation do these jobs whether or not the system is epistemically convenient. Epistemic convenience matters

---

[17] The experiment does not allow us to tell what happens for any two specific situations (i.e., any specific choice of values for the u_j) but only certain coarser facts. For instance, if u_1 is causally unanimous across *all* situations (i.e., it is either causally positive across all, or causally negative or causally neutral), for a two-valued outcome, x_e, it can be shown that the probability of x_e in the treatment group is respectively greater than, less than or equal to that in the control group iff u_1 is causally positive, negative or neutral with respect to x_e.

because we were trying to find out, not about the effects of u_1, but rather about the effects of x_1. In the case I described above, where epistemic convenience fails, u_1 has *multiple capacities*: it can affect x_e differently by different routes. We are interested only in its effect via x_1, which we shall use to calculate the effect of x_1 itself.[18] Randomization does not help with this problem. Just as in the brute force application of concomitant variation, we need either to find a cause of x_1 which we know has no other way of affecting x_e, or we need to know the overall effect via other routes in order to subtract it away.

The placebo effect is a well-known example of this problem. Getting the experimental subjects to take the medicine not only causes them to have the medicine in their bodies. It can also affect recovery by producing various psychological effects—feeling cared for, optimism about recovery, etc.

This is a good example to reflect on with respect to the general question of how widespread epistemically convenient systems are. How do we canonically deal with the placebo effect? We give the patients in the control group some treatment that is outwardly as similar to the treatment under test as possible but that is known to have no effect on the outcome under study.

That is, we do not hunt for yet another way to get the medicine into the subjects, a way that does not affect recovery by any other route. Rather we accept that our methods of so doing may affect recovery in the way suggested (or by still other routes) and introduce another factor into the control group that we hope will just balance whatever these effects (if any) may be. Ironically then, the standard procedure in these medical experiments does not support the claim that there is always a way to manipulate the cause we want to test without in any other way affecting the outcome. Epistemic convenience definitely makes randomized treatment/control experiments easier, but there are vast numbers of cases in which we do not rely on it to hold.

6 Motivations for Epistemic Convenience: 'Excessive' Pragmatism

This is a hypothesis raised by the students in my Ph.D. seminar on causality in economics at LSE: Advocates of modularity elevate a plausible answer to the question "Of what use to us is a concept of causation?" into a truth condition. This motivation is explicitly acknowledged by Daniel Hausman

---

[18] It may be just worth reminding ourselves, so as not to confuse the two issues, that x_1 itself may have multiple capacities with respect to x_e. Simple randomized treatment / control experiments do not disentangle these various capacities but rather teach us about the overall effect of x_1 on x_e.

(1998, 96-97) in defense of a similar condition to the one we are investigating:

> What do people need causal explanations or a notion of causation for? Why isn't it enough to know the lawlike relations among types? Because human beings are actors, not just spectators. Knowledge of laws will guide the expectation of spectators, but it does not tell actors what will result from their interventions. The possibility of abstract intervention is essential to causation...

My remarks here are identical to those about operationalism. Whether or not we wish to adopt the pragmatic justification as a truth condition, it will not do the job of defending modularity as a truth condition. Consider the same example as above. The conditions for using variations in u_1 to produce variations in x_1 and thereby to obtain predictable variations in x_e are much the same as the conditions for testing via concomitant variation.

To know what we will bring about in x_e by manipulating u_1 it is not enough to know just the influence of u_1 on x_1 and of x_1 on x_e. We also need to know the overall influence of u_1 on x_e by all other routes. Knowing that the size of influence by other routes is zero is just a special case. Whatever its value, if we know what the value is we can couple this knowledge with our knowledge of the influence of u_1 via x_1 to make reliable predictions of the consequences of our actions. So we do not need modularity to make use of our causal knowledge.

There is, however, a venerable argument for a different conclusion lurking here. If we are to use our causal knowledge of the link from x_1 to bring about values we want for x_e, it seems that some cause or other of x_1 must not itself be deterministically fixed by factors independent of our wishes: there must be some causes of x_1 that we can genuinely manipulate. But again, whether or not this is a good argument, it does not bear on modularity. To make use of our knowledge of the causal link between x_1 and x_e we may need a cause of x_1 that we can manipulate; but that does not show that we need a cause we can manipulate without in any other way affecting x_e.

7 Motivations for Epistemic Convenience: 'Excessive' Optimism

This is my hypothesis. Life becomes easier in a number of ways if the systems we study are economically convenient. Statistical inference of the strengths of coefficients in linear equations can become easier in well-known ways. So too can causal inference, in ways I have discussed here. And, as we shall see, Judea Pearl can provide a very nice semantics for counterfactuals as well as for a number of distinct causal notions. Wishful thinking can lead us to believe that all systems we encounter will meet the conditions that make life easier. But wishful thinking must be avoided here, or we will be led into the use of methods that cannot deliver what we rely on them for.

I think we can conclude from these considerations that these three motivations do not provide strong enough reason to accept universal economic convenience. What positive arguments then are on offer on its behalf?

8   For and Against Economic Convenience

8.1 Hausman's Defense

Daniel Hausman points out that the cause we focus on is not generally the complete cause. A complete cause will include both helping factors and the absence of disturbances. Even if effects share the causes we normally focus on (e.g., in the toaster in Figure 1, the breaking of the circuit and the moving of the rack are both caused by the motion of the lever), they will not share all of these other factors, Hausman maintains.

Disturbing factors. This claim seems particularly plausible with respect to disturbing factors. Most of the effects we are modeling here are fairly well separated in time and space. So it seems reasonable to expect that some things that might disturb the one would not disturb the other. This seems promising for the thesis of, if not universal, at least widespread, epistemic convenience. But there is a trouble with disturbing factors: often what they do is to disrupt the relation between the causes and the effect altogether. To salvage economic convenience, they need instead to cooperate with the causes adding or subtracting any spare influence necessary to ensure that the effect can take all the values in its allowed range. So they do not seem to satisfy reliably the conditions for epistemic convenience.

Helping factors. Return to the toaster. The motion of the lever causes the motion of the rack. That of course depends on the fact that the lever is bolted solidly to the rack: if the lever were not bolted to the rack, the lever could not move the rack. Could we not then take the fact that the lever is bolted to the rack to be just what we need for the special cause of the motion of the rack, a cause that the motion of the rack has all to itself?

I think not, for a number reasons:

1. As I urged in section 4, the fact that the two are bolted together is not one of the causes within the system of causal laws but rather part of the identification of what that systems of laws applies to, and this identification matters. We do not, after all, seek to know what the causal law is

that links the movement of levers *in general* with the movement of racks of the right shape to contain toast. Surely there is no such law. Rather we want to know the causal relation, if any, between the movement of the lever and the movement of the rack *in a toaster of this particular design*. Without a specific design under consideration, the question of the causal connection, or lack of it, between levers and racks is meaningless.

2. Let us, however, for the sake of argument, admit as a helping cause in the laws determining the motion of the rack the fact that the lever and rack are bolted together. My second worry about calling on helping factors like this to save epistemic convenience depends on the probability relations these factors must bear to each other. In section 4, I queried whether these factors would be probabilistically independent of each other. Here I want to ask a prior question. Where is the probability distribution over these factors supposed to come from and what does it mean?

We could consider as our reference class toasters meeting the specific set of design requirements under consideration. Then the probabilities for all of these "helping factors" being just as they are could be defined and would be 1. Independence would be trivially obtained, but at the cost of the kind of variation we need in the values of the u's to guarantee, via our theorems, that concomitant variation will give the right verdicts about causality.

Alternatively the reference class could be the toasters produced in a given factory following the designated design. Presumably then there would be some faults some time in affixing the lever to the rack so that not all the u's would have probability 1. But will the faults be independent? If not this reference class, then what? It will not do to have a make-believe class, for how are we to answer the question: if the attachment of the lever to the rack *were* to vary, what would happen to the rack? We need some other information to tell us that—most usually, I would suppose, knowledge of the causal connections in the toaster! And if not that information exactly, I bet it would nevertheless be information sufficient to settle the causal issue directly, without detour through concomitant variations.

3. The third worry is about the range of variation. For the theorems to guarantee the reliability of the method of concomitant variation, we need u's that will take the cause under test through its full range relative to the full range of values for the other causes. Otherwise there could be blips—the causal equation we infer is not true across all the

values but depends on the specific arrangement of values we consider. Will the factors we pick out have a reasonable range of variation? This remark applies equally well to disturbing factors.

4. Last I should like to point out two peculiarities in the way people often talk about the factors designated by the u's. Often they are supposed not only to represent the special causes peculiar to each separate effect but also all the "unknown" factors we have not included in our model. But if they are unknown, they can hardly be of use to us as handles for applying the method of concomitant variation. And if epistemically convenient systems are not going to be of epistemic convenience after all, why should we want them? I realize that the issue here is not supposed to be whether we want systems of this kind, but rather whether we have them. Still in cases like this where the answer is hard to make out, the strategy should be to ask "What depends on the answer". That is the reasonable way to establish clear criteria for whether a proffered answer is acceptable or not.

The second peculiarity arises from talking of the u's as a "switch" that turns the cause to different values. Often it is proposed that the switch is usually "off" yet could be turned on to allow us to intervene. This raises worries about the independence requirements on the u's again. Why should that kind of factor have a probability distribution at all, let alone one that renders it independent of all the other switch variables?

8.2 Pearl's Defense

Judea Pearl supposes that modularity holds in the semantics he provides for singular counterfactuals. He claims that, without modularity, counterfactuals would be ambiguous.[19] So modularity must obtain wherever counterfactuals make sense. This will double as an argument for universal modularity if we think that counterfactuals make sense in every causal system.

Pearl assumes modularity of the first kind, where alternative causal systems of just the right kind are always possible, but I can explain something of how the semantics works using the epistemically convenient systems we have been studying here. We ask, for instance, in a situation where $x_j = X_j$ and $x_k = X_k$, "Were $x_j = X_j + \Delta$, would $x_k = X_k + a_{kj}\Delta$?" The question may be thought ambiguous because we do not know what is to stay fixed as x_k varies. Not so if we adopt the analogue of Pearl's semantics for our epistmically convenient system. In that case u_j must vary in order to produce the required variation in x_j and all the others u's must stay the same.

---

[19] 14 March, 2000, seminar presentation, Dept. of Philosophy, University of California at San Diego.

The semantics Pearl offers is very nice, but I do not see how it functions as an argument that counterfactuals need modularity. The counterfactuals become unambiguous just because Pearl provides a semantics for them and because that semantics always provides a yes-no outcome. Any semantics that does this will equally make them unambiguous.

Perhaps we could argue on Pearl's behalf that his is the *right* semantics, and it is a semantics that is not available in systems that are not epistemically convenient. Against that we have all the standard arguments that counterfactuals are used in different ways, and Pearl's semantics—like others—only accounts for some of our uses. We should also point out that we do use, and seem to understand, counterfactuals in situations where it is in no way apparent that the causal laws at work are epistemically convenient.

I think one defense Pearl may have in mind concerns the connection between counterfactuals and causality. Consider a very simple case where one common cause, v, is totally responsible both for x_1 and x_2 and no u_1 is available to vary x_1 independently of v.[20]

It is easy to construct a semantics, similar indeed to the one Pearl does construct, that answers unambiguously what would happen to x_2 if x_1 were different. This semantics would dictate that we vary v to achieve the variation in x_1. Then of course x_2 would vary as well. So it would be true that were x_1 to be different, x_2 would be. And that seems a perfectly reasonable claim for some purposes. But not of course if we wish to read singular causal facts from our counterfactuals. So Pearl could argue that his semantics for counterfactuals connects singular counterfactuals and singular causal claims in the right way. And his semantics needs modularity. So modularity is a universal feature wherever singular causal claims make sense.

Laying aside tangled questions about the relations between singular causal claims and causal laws, which latter are the topic of this paper, I still do not think this argument will work.

We could admit that for an epistemically convenient system Pearl's semantics for counterfactuals plus the counterfactual-causal links he lays out will give correct judgments about causal claims. We could in addition admit that causal claims cannot be judged by this method if the system is not epistemically convenient. All this shows is that methods that are really good for making judgments in one kind of system need not work in another kind.

More strongly, we could perhaps somehow become convinced that no formal semantics for causal claims that works, as Pearl's does, by trans-

---

[20] For Pearl this would mean that there was no alternative causal system possible that substituted the law "Let $x_1 = X_1$," for the law connecting x_1 and v.

forming a test into a truth condition, will succeed across all systems of laws that are not epistemically convenient. That would not show that there are no causal laws in those systems, but merely that facts about causal laws are not reducible to facts about the outcomes of tests.

9 Conclusion

Modularity is not a universal feature of deterministic causal systems, nice as it would be were it universal. Part of my argument for this conclusion depends on asking of various factors, such as the fact that the toaster rack is bolted to the lever, "Are these really causes?" I argued that they are not because they cannot do for us what we want these particular kinds of causes to do. In this case what we want is a guarantee that if we use these factors in applying the method of concomitant variation, the results will be reliable.

I think this is the right way to answer the question. We should not sit and dispute whether a certain factor in a given situation is really a cause, or what causality really is. Rather we should look to whether the factor will serve the purposes for which we need a "cause" on this occasion. That means, however, that for different purposes the very same factor functioning in the very same way in the very same context will sometimes be a cause and sometimes not.

That is all to the good. Causality is a loose cluster concept. We can say causes bring about their effects, but there are a thousand and one different roles one factor can play in bringing about another. Some may be fairly standard from case to case; others, peculiar to specific structures in specific situations. Causal judgements, and the methods for making them reliably, depend on the use to which the judgment will be put. I would not, of course, want to deny that there may be some ranges of cases and some ranges of circumstances where a single off-the-shelf concept of causality, or a single off-the-shelf method, will suffice. But even then, before we invest heavily in any consequences of our judgments, we need strong reassurance both that this claim is true for the ranges supposed and that our case sits squarely within those ranges.

That of course makes life far more difficult than a once-and-for-all judgment, a multipurpose tool that can be carried around from case to case, a tool that needs little knowledge of the local scene or the local needs to apply. But it would be foolhardy to suppose that the easy tool or the cheap tool or the tool we happen to have at hand must be the reliable tool.

References

Cartwright, N. (1989) *Nature's Capacities and Their Measurement.* Oxford: Oxford University Press.

Cartwright, N. (1999) *The Dappled World: A Study of the Boundaries of Science.* Cambridge: Cambridge University Press.

Cartwright, N. (2000) *Measuring Causes: Invariance, Modularity and the Causal Markov Condition.* London: CPNSS Measurement in Physics and Economics Discussion Paper Series, LSE.

Cooley T. F. and LeRoy, S. F. (1985) A theoretical macroeconometrics. *Journal of Monetary Economics,* 16, 283-308.

Hausman, D. (1998) *Causal Asymmetries.* Cambridge: Cambridge University Press.

Hausman, D. and Woodward, J. (1999) Independence, invariance and the causal Markov condition. *British Journal for the Philosophy of Science,* 50, 1-63.

Hoover (forthcoming) *Causality in Macroeconomics.* Cambridge: Cambridge University Press.

Macaulay, D. (1988) *The Way Things Work.* Boston: Houghton Mifflin.

Pearl, J. (2000) *Causality.* Cambridge: Cambridge University Press.

Simon, H. (1977) Causal ordering and identifiability. In H. Simon (Ed.) *Models of Discovery and Other Topics in the Methods of Science.* Boston: Reidel, 53-105.

Woodward, J. (1997) Explanation, invariance and intervention. *British Journal for the Philosophy of Science* 64, 26-41.

Woodward, J. (2000) Causation and manipulation. ms, California Institute of Technology.

5

Three Dogmas of Humean Causation

GÜROL IRZIK

Boğaziçi University

1 Introduction

Dogmas are beliefs held come what may. They survive despite contrary evidence and argument. As I see it, Humean causation is a dogmatic doctrine that consists of the following:

The First Dogma: Ontological Reduction. Exceptionless regularities or probabilistic-correlational complexes are constitutive of causal relations. The latter are reducible to the former.

The Second Dogma: Supervenience. Singular causal facts are true by virtue of generic causal facts (causal law statements).

The Third Dogma: Epistemological Reduction. Our knowledge of causal relations (generic or singular) is based on our knowledge of regularities, or probabilistic-correlational complexes, and of spatio-temporal relations. This is essentially an inductive, therefore, inferential kind of knowledge.

Leaving aside the part about probabilities and correlations, we owe these dogmas to David Hume (1739), so they have been with us for a long time. The qualifications concerning exceptionless regularities were added much later (in 1970, to be exact, by Patrick Suppes) in order to avoid some counter arguments and examples. This marks the beginning of the end of the Humean hegemony in matters of causality, as alternative approaches have proliferated

Stochastic Causality.
Maria Carla Galavotti, Patrick Suppes and Domenico Costantini (eds.).
Copyright © 2001, CSLI Publications.

since then. To name a few, we now have the process approach, the capacity approach, the DAG approach, the agency approach, and a number of probabilistic approaches to causality. This boom, I believe, is due in part to the rediscovery by philosophers of new tools like conditional probabilities and in part to a liberation from the first dogma; all of these new approaches employ probabilistic or statistical apparatuses one way or another, and all contain strongly anti-reductionist elements. One can even show that the two reasons are related in interesting ways, but that is not the purpose of my paper. What I instead would like to argue is that while recent accounts of causality have gone a long way toward emancipation from the grip of Humeanism, one last vestige still remains. This is the third dogma, which in effect says that no causal relation is observable and that none of our causal knowledge is innate. It has received surprisingly little critical attention from philosophers, even from those who reject the first two dogmas. Inspired by the pioneering work of Albert Michotte, psychologists have been for years carrying out experiments that strongly suggest that even very young children can perceive causal relations. Philosophers of diverse inclinations persistently ignored and continue to ignore such work. What better label than "dogma", then, can characterize the thesis of epistemological reduction?

The plan of my paper is as follows. In the next section I will summarize the main objections to the first dogma, that is, the thesis of ontological reduction. I will say very little about the second dogma, but the interested reader can consult Cartwright (1989) and Woodward (1990). Section 3 will be devoted to some aspects of the capacity approach, particularly in the form developed by Nancy Cartwright. I believe that the capacity approach provides the most promising framework to make sense of causation. However, it faces an hitherto unnoticed epistemological problem, a circularity concerning our knowledge of causal claims understood as ascription of capacities. I will introduce this problem and offer a way of solving it. Although my solution has restricted scope, it seems to me to be fruitful. For by involving precisely the rejection of the third dogma, it opens up new avenues for investigation. Naturally, the bulk of my paper will be devoted to defending, in sections 4 and 5, the view that at least some causal relations are observable. My defense will be based on both empirical and philosophical grounds. I will argue that the third dogma of Humean causation results from a very poor epistemological theory; that there exists a much better philosophical account of seeing and knowing, which allows us to say that causal relations can be observed; and, finally, that there is considerable empirical evidence supporting the observability of some causal relations.

2 The First Dogma: Virtues and Vices

It is not difficult to see why a Humean account of causality incorporating the three dogmas outlined above appeals to so many philosophers: it has an anti-metaphysical, deflationist-reductionist ontology and has an attractive "solution" to the epistemological problem. Regularities conveniently do both ontological and epistemological work. These virtues are nicely displayed by Hume's own definition: "We may define a cause to be an object precedent and contiguous to another, and where all the objects resembling the former are placed in like relations of precedency and contiguity to those objects that resemble the latter" (Hume 1739/1962, p. 221). To put it differently, an event A causes another event B if and only if A is earlier than B, A is contiguous to B, and whenever events of type A occur events of type B follow. Ontologically, this definition employs only spatiotemporal relations which exhibit a regular, law-like behavior. A world that contains regularities automatically contains causal relations as well simply because regularities are constitutive of causal relations. As causal relations are reducible to constant conjunctions, the Humean world has no ontological excess like powers, natures, or necessities for connecting causes to their effects. Such a world is presumably more easily accessible because all one needs in order to have knowledge of causes is knowledge of regularities, and that can be acquired by induction over observations of repeated instances of A and B.

Despite its virtues, however, Hume's definition has a number of well-known vices. It is worth recalling some of them. First, there are constantly conjoined pairs of events which are not causally related. EPR-type correlations are a case in point. To take a pedantic example from daily life, nights follow days regularly, but one is not the cause of the other. Hume's view has no way of distinguishing between genuine causal relations and 'spurious' constant conjunctions symptomatic of common causes. Second, not all causal relations imply exceptionless regularities; some are probabilistic, and Hume's definition has no resources to cope with them. Finally, there seems to be causal relations which do not display any regular behavior. Here is a typical example, which I believe I read in one of Elizabeth Anscombe's papers: my stepping on ice (call it C) made me fall down (call it E) the other day, though it is not the case that each time I step on ice I fall down. So, there does not seem to be any law that relates stepping on ice to falling. Donald Davidson's ingenious reply has become the standard one to all such counterexamples: The law does not have to connect the cause to its effect under those descriptions. Rather, C falls under a certain event kind or description c, E falls under a certain event kind or description e, and the law relates c to e. Davidson claims that such a law always exists: "In any case, in order to know that a singular causal statement is true, it is not necessary to know the truth of the law; it is necessary only to know that some law covering the events at hand exists" (Davidson 1968, pp. 93-94). Now, what reason is there to think that some such law always exists? Davidson writes: "And very often, I think, our justification for accepting a singular causal statement is that we have reason to believe an appropriate causal law exists, though we do not know what it is." (Davidson 1967, p. 701). There are two difficulties with this sort of reasoning. First, covering laws are sparse. Second, the postulation of the existence of a covering law is not the only justification we have for accepting a singular causal statement. Often we rely on the elimination of other possible causes. Thus, if we doubt whether it was A that caused B, we carefully review the circumstances to see if some factor other than A could have produced B. Failure to find it justifies the conclusion that A caused B. The Humean account blinds us to the eliminationist strategy which is at the heart of all experimental design and reasoning.

It is mainly for these reasons that philosophers have turned to probability theory to rescue Hume's program. A notable attempt is Patrick Suppes' *A Probabilistic Theory of Causation* (1970). According to his account, effects need not follow their causes regularly; they just need to occur more often than not when their causes are present. This is usually expressed with the slogan that a cause must increase the probability of its effect. More precisely, an earlier event A is said to cause a later event B if and only if $P(B/A) > P(B)$ and there exists no event C earlier than both A and B, which screens off A from B. C is said to screen off A from B if and only if $P(B/A) > P(B)$ and $P(B/A.C) = P(B/C)$.

The idea of screening off turned out to be fruitful indeed, especially in capturing 'spurious' correlations arising from common causes: a common cause screens off the correlation between its effects. Furthermore, screening off is an asymmetric relation in the sense that if C screens off A from B, A does not necessarily screen off C from B. This was encouraging because causal relation is also asymmetric, so there was reason to hope that the notion of screening off, together with the idea that a cause raises the probability of its effect, may provide a reductive definition of causation after all.

Unfortunately, as we now know very well, not only common causes but also intermediate causes function as screeners off. The standard way to distinguish between these cases is to consider time. However, some philosophers are reluctant to appeal to temporal order and direction, which, they believe, arise from causal direction and order (see, for example, Mackie 1980, Papineau 1985), so for them reduction is even more elusive.

Intermediate causes are not the only kind of problem that stands in the way of reduction even when temporal considerations are taken into account. Simpson's 'paradox' is another. Suppose that A causes B and that there is no further factor which screens off A from B. So we expect the probability of B given A to be greater than the probability of B in general. Assume furthermore that A also contributes to C, which prevents B. In that case, A is a positive causal factor for B through one path and a negative causal factor through another. If these two influences cancel each other out, we may not observe the expected probability increase.

Having been convinced that these difficulties are insurmountable, a number of philosophers rejected the first dogma; that is, they gave up the project of reducing causal relations to regularities or probabilistic dependencies and turned to the less ambitious but perhaps more fruitful task of establishing the right kind of connection between them. Salmon's process approach and Cartwright's capacity approach are well-known examples. More recently, Spirtes, Glymour and Scheines (1993), Pearl and Verma (1994), who are the main founders of the DAG approach, have joined the non-reductionist camp.

However, some philosophers, most notably David Papineau (1991, 1993) and Wolfgang Spohn (this volume), believe that the DAG approach can be used for reductive purposes. They employ very similar strategies, the major difference being that Papineau attempts to define not only causal relatedness but also causal directionality in terms of the resources of the DAG approach. I have argued against his version in detail elsewhere (see Irzik 1996). So, let me turn to Spohn's. The key idea is that causal dependence relations among a given set of variables U (called the frame) can be represented by a directed acyclic graph (called DAG). A probability measure can be assigned over the members of U, so there are probabilistic dependence and independence relations among them. Basically, three conditions specify the kind of relations that hold between the frame U represented by a DAG and the associated probability measure. The first is *the Markov condition* that says that each variable is probabilistically independent of all its non-descendants conditional on its parents. Screening off relation defined above is a special case of this. The second condition says that no proper subgraph of the DAG satisfies the Markov condition. This is called *the minimality condition*. The third condition is known as *the faithfulness condition*. Its exact formulation is a bit complicated, but its intuitive content is clear: no conditional independence relation (such as screening off or a zero partial correlation) in U results from accidental situations regarding the parameter values of U. In other words, every such relation is determined by the structure of the DAG alone. A DAG that satisfies all three conditions is called a *Bayesian net*. Spohn then asserts that every causal graph is a Bayesian net and defines causal dependence in terms of it: B directly causally depends on A if and only if there is a causal path from A to B in a Bayesian net (Spohn, this volume). In other words, Bayesian nets are all there is to causal relatedness.

But are these conditions universally satisfied? Take the faithfulness condition, for example. All cases of Simpson's 'paradox' are violations of this condition; whenever we have a cause that acts as a contributor through one route and that acts as a preventer through another in such the way that the two influences cancel each other out, the faithfulness condition will be violated. Admittedly, such cases will be rare, so methodologically it may be all right to impose it on our models, but it will not do if the aim is ontological reduction. Violations of the faithfulness condition, no matter how rare they are, form a serious obstacle for reductionist accounts such as Spohn's and Papineau's.

The Markov condition does not fare better in this regard either. For one thing, there is the EPR paradox that violates it. For another, as Nancy Cartwright has shown, the condition also fails when a cause acts in a truly probabilistic way. To see this, consider the following abstract example provided by her (Cartwright 1999, p. 109; see p. 108 for a concrete example). Suppose there is a cause C that has two separate yes-no effects, X and Y. We then have four possible outcomes: +X+Y, -X+Y, +X-Y, -X-Y. Assuming that C occurs, there are four joint probabilities that must be fixed by the world: Prob(+X+Y), Prob(-X+Y), Prob(+X-Y), Prob(-X-Y). As Cartwright puts it, "nothing in the concept of causality, or of probabilistic causality, constrains how Nature must proceed" (ibid.). The Markov condition holds only in one very special case, where Prob(+X+Y).Prob(-X-Y) = Prob(+X-Y).Prob(-X+Y). In all other cases, it will fail. Short of determinism, then, there is no reason to expect that the Markov condition will be satisfied universally. As before, this does not mean that Markov condition is useless methodologically, but it does imply that ontological reduction does not succeed.

The typical reductionist response to such objections is to argue that it is always possible to find a larger, more refined graph-theoretic frame in which the Markov, the minimality, and the faithfulness conditions are satisfied. This is the strategy pursued by David Papineau, for example. He claims that a world of cosmic conspiracies in which the probabilistic dependencies and independencies conceal the true causal structure of the world forever is conceivable but metaphysically impossible (Papineau 1993, p. 246). In a similar vein, Spohn claims that "in the final analysis it is the all-embracive Bayesian net representing the whole of reality which decides about how the causal dependencies actually are" (Spohn, this volume). Spohn denies that we possess an independent concept of causality to rely on, independent, that is, from Bayesian nets.

Such strong claims need some reason for believing them! I think it is plain false that we do not have a notion of causal relatedness independent of Bayesian nets. Our basic notion of causality is simply that causes make their effects happen, and I claim that we acquire this notion primarily by manipulating objects; in other words, it is ultimately tied to the fact that we are agents. I will say more about this in section 6. Let us first look at what happens to the reductionist project with Papineau's and Spohn's moves. That project was attractive to empiricist philosophers because it was thought to be anti-metaphysical and epistemologically feasible. Now, it has itself become metaphysical and epistemologically elusive. To be consistent, reductionists *must* deny the possibility of a "non-Bayesian" or "conspiratorial" world. They appeal to probabilistic relations because they believe that they are empirically more easily accessible, and they reject notions such as power and capacity because they find them metaphysical. To accept that probabilities may never reveal the true causal structures which they constitute would be of course self-defeating. But the commitment to the existence of a "non-conspiratorial" causal world (Papineau) or an "all-embracive Bayesian net representing the whole of reality" (Spohn) surely transcends all possible experience since only God can have this sort of knowledge. Consequently, such a commitment is itself metaphysical, and the first dogma survives only within this deadly dialectic.

3   An Alternative Approach: Causes as Stable Capacities

In my view, the starting point for an alternative approach should be to accept without embarrassment that causation is a primitive relation that cannot be defined in terms of non-causal relations. The notion of cause makes sense only within a semantic field of related notions such as capacity, nature, change, manipulation, and invariance. I do not believe that these terms can be defined independently from each other, nor do I believe that they can be learned in isolation from one another; they must be acquired together in clusters. Thus to know the meaning of 'cause' is to know the network of relations in the semantic field to which it belongs. I am, in other words, advocating a semantic holism with respect to the concept of cause.

Take a causal claim like "smoking causes lung cancer". Following Cartwright (1989, 1999) and Harre and Madden (1975) I suggest that it should be understood as an ascription of a capacity or power to smoking to produce lung cancer. To say that X has the capacity or power to Y means that "X can do Y, in the appropriate conditions, *in virtue of its intrinsic nature*" (Harre and Madden 1975, p. 86; emphasis original). Thus, there is something in the nature of smoking, probably certain chemicals, the inhalation of which can produce lung cancer under certain conditions. Often, these conditions must be created artificially, in labs. The capacity or power in question is a more or less stable one that is manifested in different situations, "a capacity which if the circumstances are right reveals itself producing a regularity, but which is just as surely seen in one good single case" (Cartwright 1989, p. 3). In this account regularities are no more than empirical manifestations of underlying nature of things under very special circumstances.

The notion of stable capacity is tied to the notion of invariance. Causal claims must satisfy an invariance condition in the sense that the relation between a cause and its effect should continue to hold under some specified class of changes in various conditions including initial and background ones. In the case of smoking-lung cancer, for example, the causal relation should continue to hold under changes of age, sex and the like.

It may also be worth noting that 'capacity' is clearly an Aristotelian notion. Aristotle's term was potency, by which he meant "the source of change or movement in another thing, being moved by another thing or by itself qua other" (Metaphysica, Book Δ, ch. 12, 1019a). So, the notion of change is built into the notion of cause. If A and B are causally related, manipulating A results in a change in B. That is why quitting smoking, but not cleaning one's yellow fingers, is an effective strategy to reduce the chance of having lung cancer, despite the fact that having yellow fingers and having lung cancer are highly correlated.

Roughly, this is the semantic holism I have had in mind. Although I have not defended the capacity approach, I believe that it provides us with the right framework to make sense of causal claims both in the scientific and ordinary contexts. The interested reader should consult the excellent works of Cartwright (1989, 1999). What I want to do in this section is to draw attention to a problem that faces the capacity approach, a problem that threatens it with an epistemological circularity.

To see this, begin by noting the distinction between causal capacity claims and singular causal claims. For the latter to be true, the cause and the effect must actually occur. But a causal capacity claim can be true even if the cause and the effect in question never occur. For example, after the devastating earthquake in the Marmara Region of Turkey in 1999 so much stress has been put on the fault line near Prince Islands that it now has a capacity to produce a major earthquake in Istanbul, even though it has not happened yet. There is a straightforward relationship between singular causal and causal capacity claims: for a singular causal claim to be true, it is necessary that a corresponding causal capacity claim must also be true. By contrast, the truth of a singular causal claim is sufficient for the truth of a capacity claim.

Let us then ask how we know the truth of a singular claim of the form "A caused B". Well, to know this we must know that A is the sort of thing that can cause B; if A does not have the right nature and therefore the right capacity, it cannot produce B whatever the circumstances are. But how do we know that A has that sort of capacity? If the answer is, only by knowing that it or things similar to it have exercised that kind of capacity in various situations, i.e., by knowing the truth of some appropriate singular causal claims, then we are running in a circle.

The circle may not be a vicious one. We know that the fault line under Prince Islands has the power to produce a major earthquake, though it has not exercised its power yet. We know this because we know that that fault line is an extension of the North Anatolian fault line, which did produce similar earthquakes elsewhere in the past. Although what we are facing here is not a vicious circle, it is nevertheless worrisome. For the problem still remains: how does our knowledge of causes understood in terms of capacities get off the ground? After all, token causal relations are the result of exercising of capacities, and capacities are not open to direct inspection.

It seems to me that a profitable way out of this circle involves the denial of the third dogma that our knowledge of causal relations is always inferentially obtained by induction. This opens up two possibilities: some of our causal knowledge may be innate, or it could be observational. Recent work in developmental psychology suggests both. Let me start with innateness.

According to what is called "the theory theory", "the processes of cognitive development in children are similar to, indeed perhaps even identical with, the processes of cognitive development of scientists" (Gopnik and Meltzhoff 1997, p. 3). Just as scientists have theories about the world, which they revise according to evidence they gather, so do infants. To put it differently, human beings are born with certain substantive principles about the world and mechanisms for revising them on the basis of their experiences. For example, there is considerable evidence that infants have an innate knowledge that the world is three-dimensional and that objects move in certain trajectories (ibid., chs. 4 and 5). In a similar vein, Gopnik et al. (forthcoming) argue that the process of intuitive 'theory' formulation and revision also involves a type of representation that they call a causal map. They define a causal map as an abstract representation of the causal relationships among events in the world. They propose that certain cognitive devices were designed by evolution to recover causal information, and that animals, including human beings, may have some hard-wired expectations about possible causes and causal behavior. These constrain the attention of the infants to only certain aspects of the world. In the language of the capacity approach, this means that human beings from birth have some knowledge about capacities of objects. Needless to say, this is a very limited, general and abstract sort of knowledge, but, contra Humeans, it is innate. Experience serves only to confirm or revise it. This is a non-inductive, a top-down kind of knowledge, so to speak, and breaks the circle we mentioned earlier: some knowledge of capacities is given innately without prior knowledge of any singular causal truths.

This is the ground base from which all our knowledge of causes takes off, a base which also makes possible a bottom-up kind of causal knowledge when coupled with observation. In other words, once we have knowledge of capacities, in certain circumstances we can observe one event causing another, and that gives us knowledge of singular causal truths without appealing to any knowledge of corresponding regularities. As we shall see, developmental psychology provides considerable empirical evidence for this as well. I do not mean to suggest that all our causal knowledge is acquired in this way, nor do I deny that many casual relations are unobservable. What I claim is that if we can see tables, fires, and so on, we can also see a hammer's smashing a tomato, a boy's pushing another boy, and so on. I will present the findings of the developmental psychologists in section 5. Let me now discuss some likely objections and then provide an analysis of what it is to see a causal relation.

4   Observability of Causal Relations

The idea that causal relations can be observed would be appalling to a Humean. I suppose he would object as follows. For one thing, he would say that all we observe is one event being followed by another: we observe first the blow of the hammer and then the smashed tomato, but never one causing the other, much less the relevant capacity. For another, he would ask us how we know that it is the blow of the hammer that smashed the tomato without establishing a regular occurrence between the two. Indeed, the two objections are related in that an archaic skepticism lurks behind them. To the first objection I reply: We do not see the powers, we infer them. They come with the properties that enter into causal relations. Just like there are observable objects, events and properties, there are observable relations in the world as well. If being taller than is an observable relation, I do not see why smashing, pulling, pushing, grabbing and a host of similar relations cannot be observed as well. From this viewpoint, there is nothing peculiar about causal relations. Perceptual skepticism might have made sense at the time of Hume who believed that all we perceive are our immediate sensations of sight, touch, pain, etc., but it does not make sense today. My reply to the second objection is that of course we can be mistaken about our causal judgments based on perception, but that this sort of fallibility pervades all other contexts: my observations can fail me about any other objective relation. Checking regular occurrence is surely one way of avoiding making mistakes, but it is not the only one. As we saw earlier, elimination of other possible causes is another.

Dispelling the objections of Humeans is not enough, of course. I owe a positive account of what it is to see a causal relation. Luckily, such an account has already been developed and defended in great detail by Fred Dretske more than forty years ago, and it is a pity that it has not been appreciated sufficiently by philosophers of science. In his *Seeing and Knowing* Dretske (1969) distinguishes between epistemic and non-epistemic senses of seeing. He defines the latter as follows: "S sees$_n$ D = D is visually differentiated from its immediate environment by S" (Dretske 1969, p. 20). Here, S is a subject, D is an object or event, and the subscript indicates the non-epistemic sense of seeing. 'Visually differentiated' involves, among other things, "S's differentiation of D is constituted by D's looking some way to S, and, moreover, looking different than its immediate environment" (ibid.). Non-epistemic seeing is a kind of seeing which is devoid of any positive belief content. A person can see an object in this sense without knowing what it is, without even knowing that it is an object. Thus, non-epistemic seeing is different from, and indeed possible without, 'seeing as'.

Although we can see things around us without recognizing what they are, our vision has also epistemic import. We can see, for instance, not only water but also that it is water. Dretske states four conditions that must be met in order for seeing to be epistemically informative: S sees that b is P in the epistemic way if (1) "b is P", (2) "S sees$_n$ b", (3) "The conditions under which S sees$_n$ b are such that b would not look, L, the way it now looks to S unless it was P", (4) "S, believing the conditions are as described in (3), takes b to be P" (ibid., pp. 79-88).

Once we are given a definition of epistemic seeing for objects and events, it can easily be extended to cover any relation at all. Let me apply the definition Dretske provides for relations to the relation of causality (see Dretske 1969, p. 141):

S sees that A causes (or better: is causing) B in the epistemic way if

(i)   A is causing B

(ii) S sees$_n$ A and S sees$_n$ B

(iii) The conditions are such that A and B would not look the way they do, L, relative to one another to S unless A were causing B,

(iv) S, believing the conditions are as described in (iii), takes A to be the cause of B.

Go back now to our example of the hammer and the tomato: We hit a tomato with a hammer (A) and then see the tomato being smashed (B). Now, do we also see that A caused B? The answer is an unambiguous 'yes'. Conditions (i) and (iv) pose no problem. Condition ii is satisfied because both the hammer blow and the smashing of the tomato are observable in the sense that we can distinguish them from their environment by the way they look to us. Whether condition iii is satisfied or not is an empirical matter; if the blow and the smash would not look to us the way they do were it not the case that one was causing the other, then the condition would be fulfilled. Note that it is perfectly possible that some other event C can cause the same effect without changing the way A and B would look to us. This would be the case, for instance, if a small bomb implanted inside the tomato exploded at the right moment and in the right way. In that case, condition iii would be violated, and we would not be able to see that A was causing B (cf. Dretske 1969, p. 231).

I submit that this analysis of epistemic and non-epistemic seeing provides a much more satisfactory framework than Hume's simplistic theory of impressions and ideas. Moreover, with some qualifications, it can be extended to cover touching, smelling, and so on although I lack the space here to do so. Its chief virtues for the purposes of the present paper are that it dispenses with the third dogma of Humean causation and that it is consistent with the psychologists' empirical findings to which I now return.

5 Evidence for Infants' Perception of Causal Relations

Recent research in developmental-cognitive psychology has provided considerable evidence to the effect that a causal relation can be directly observed as being distinct from the spatiotemporal properties of events. The pioneer in this field is Albert Michotte (1963) who suggested that even infants might have a direct perception of some cause-effect relationships as a kind of perceptual gestalt. Since then a number of psychologists followed up Michotte's suggestion and devised various experiments to test it. Here I will summarize the striking results of a series of such experiments conducted by Alan Leslie (1982, 1984), and Alan Leslie and Stephanie Keeble (1987). Like most others, theirs too use the habituation-dishabituation of looking technique. Six-month old infants are shown films of a red object colliding with a green object in a variety of ways. Habituation involves subjecting the infants to the image of the same motion many times. This causes a decrease in the infant's attention, which can then be measured in terms of the time of looking. Dishabituation involves presenting a contrasting image just once and then observing the infant's response in terms of the recovery of attention manifested as a longer period of looking. The idea behind the habituation-dishabituation technique is that because we are surprised at the events that violate our expectations, we look at them for a longer period of time. If this logic is correct, then subjects should be surprised at a violation on the first trial (dishabituation) and should look at this violation for longer than presentation of an event that is not in violation.

An initial experiment suggested that infants can distinguish between direct launching and similar but discontinuous events, where direct launching means that an object (here, a red one) moves continuously in a straight line to the right and collides with another object (a green one), causing it to move in the same manner (Leslie 1982). This is a perfect example of seeing in the non-epistemic sense. The experiment provides nice evidence that not only adults but also infants as young as six-month old can see$_n$ certain events.

In another experiment, the effect of direct launching was compared to the continuous motion of a single object (Leslie 1984). One group of infants was habituated to direct launching and then subjected to its reversal once, i.e., dishabituated. Another group was habituated to continuous motion of a single object and then dishabituated in the same way. The result was that the first group recovered their looking more than the second group. This suggested that the infants could detect even an internal structure and parse the submovements in direct launching.

A third experiment was designed to show that the infants could perceive something more than the spatiotemporal dimension in the connection between the events that make up direct launching (Leslie and Keeble 1987). To this end, a group of infants were habituated to direct launching and then subjected to the image of delayed reaction-without collision. Another group was habituated to launching-without-collision and then dishabituated with an image of delayed reaction. In both cases, spatial and temporal contrast between the image of habituation and that of dishabituation was equal. However, the infants in the first group showed a greater degree of recovery of attention, suggesting that they had perceived a greater change in the replacement of direct launching, which is the only apparently causal sequence, with delayed reaction-without-collision, which appears non-causal.

In a final and more conclusive experiment, Leslie and Keeble compared the reversal of direct launching with the reversal of delayed reaction (ibid.). The first group of infants was habituated to a sequence in which a red object directly launched a green object by colliding with it in a rightwards direction. They were then shown a reversal of this motion where the green object came back and directly launched the red object in a leftward direction. The second group was subjected to a similar reversal in the case of delayed reaction. Again, the first group displayed a significantly greater recovery of attention, indicating that the infants in this group were responding to the reversal of causal direction as well as that of spatiotemporal direction while the infants in the second group were responding only to the spatiotemporal reversal.

These experiments provide evidence that even very young children can perceive some of the causal relations in the world. In all likelihood, there is a causal percept factor at work in infants' observations of motion. Leslie and Keeble hypothesize that there is a visual mechanism, already operating at the age of six months, which is responsible for organizing a causal percept. They conclude that "instead of causality being entirely a result of the gradual development of thought (Piaget) or of prolonged experience (Hume), an important and perhaps crucial contribution is made by the operation of a fairly low level perceptual mechanism" (ibid., p. 285).

This is perhaps to be expected. After all, causal information is vitally important for survival and adaptation. It is therefore not surprising that human beings (and also some animals) have biologically evolved in such a way that they are endowed with certain mechanisms that specifically target the causal structure of the world.

Now, how does all this fit into the analytic framework of seeing that we have discussed earlier? I think it fits perfectly well. First of all, the experimental evidence suggests that even infants can see certain objects, events and causal relations. Of course, initially, this must be a seeing in the non-epistemic sense. Later, when they are old enough, their visual experiences begin to acquire epistemic import. By the time they are ten months old, for example, they realize the significance of spatial contact for causal efficacy. In other words, they begin to see that they can move an object only by coming into spatial contact with it and use this sort of information to make predictions (Gopnik and Meltzoff 1997, p. 138). It is probably from this point onward that non-epistemic seeing becomes epistemically clothed.

6   Concluding Remarks

Since Hume's publication of *Treatise* in 1739, the regularity account of causation has dominated the philosophical scene despite repeated failures to reduce causation to exceptionless regularities, probabilistic dependencies or correlational complexes. In the last several decades, however, what I have called the first dogma of Humean causation lost its grip, and, as a result, a number of non-reductive approaches have emerged. Some philosophers of science still hope to rescue the reductionist project, but I have argued that their attempts too fail, this time in a self-defeating manner. The rescue attempt itself becomes metaphysical, inconsistent with the original intent.

Although regularities are no more believed to be ontologically privileged, they continue to play an epistemologically privileged role in many accounts. Probabilistic or otherwise, they are often taken to be our only source of causal information, our only justification for causal claims. While I do not deny their epistemological role, I have argued that that role is limited.

We have other means of epistemological access to causes, such as the method of elimination and direct observation. I have borrowed Dretske's analysis of seeing to specify the conditions under which a person can be said to observe a causal relation. This is a far better account than Hume's which makes the observability of causal relations impossible. I have then drawn attention to recent empirical studies that provide evidence against the third dogma. These studies also suggest that, contra Humeans, some of our causal knowledge is inborn. I have used these alternative sources of causal knowledge to solve the circularity problem faced by the capacity approach.

Let me conclude by saying a few words about our concept of cause and its acquisition. As Cartwright has pointed out, 'cause' does not seem to be a unitary notion (Cartwright, this volume; see also her 1999, pp. 118-121). It is highly abstract, unspecific and varied: causes may be "standing conditions, auxiliary conditions, precipitating conditions, agents, interventions, contraventions, modifications, contributory factors, ...etc." (ibid.). This may explain why our attempts to define causation fail; there is simply no single notion of cause to be defined. Perhaps that is also why we must be semantic holists in the sense I explained in section 3. To know the meaning of the term 'cause', we must know the meanings of a cluster of related terms like 'change', 'produce', 'manipulation', 'intervention', 'prevention', 'power', and the like.

How do we then acquire such a non-unitary notion? The answer, I believe, is by extension from the primordial cases. By primordial cases I mean our earliest experiences as agents. Again, developmental psychology provides some clues (see Leslie and Keeble 1997, ch. 5). Apparently, infants can distinguish people from inanimate objects very early on and become aware of the effects of their actions on such objects when they are only two or three months old. That is to say, they seem to notice the causal connection between their actions and their effects. The paradigmatic examples are grabbing a toy, sucking a pacifier, pulling a napkin, pushing a cup, and so on. This awareness in turn enables them to make predictions about which events will be followed by which actions. At this early stage, however, infants' attention is directed to the temporal dimension of the link between their actions and their consequences. Around the age of ten months, they begin to realize the importance of the spatial contact between the two. For instance, at this age they know not only that they can move objects by kicking them, but also that stronger kicks result in greater displacement.

A second primordial source of our notion of cause is the link between mental states and actions. This is established primarily through imitation. There is overwhelming evidence that infants who are only a few weeks old can imitate facial gestures such as tongue protrusion and mouth opening.

Since such imitations "require a mapping from visually perceived physical movements to internally felt kinesthetic sensations, the most fundamental of action representations," and since such sensations are closely related to mental states such as pain, there seems to be an innately established connection between these internal sensations and facial expressions (ibid., p. 130). "Similarly", write Leslie and Keeble, "simple motor plans, like the intention to move your tongue, seem to be a primitive kind of mental state, and we seem to map these plans onto perceived actions" (ibid., p. 131).

It seems to me that it is these kinds of experiences that constitute the primordial sources from which we acquire our initial concept of cause. Since it essentially derives from our actions as agents, it involves 'change', 'make happen' and 'bring about', which is the standard dictionary meaning. Once we have this 'core' meaning, we extend it to situations that are sufficiently similar to the original ones: a bee's sting, objects colliding, wind blowing leaves, fire burning paper, and so on. How we extrapolate to such cases from the 'core', which aspects of objects, their properties and relations we pick out as similar, of course, requires careful empirical scrutiny. This is all the more so in scientific contexts. One rule of thumb for extension seems to be manipulability. Anything that can be used to manipulate anything else for a certain purpose can be classified as a cause. Perhaps regularity and counterfactual dependence constitute other such rules.

This account accords well with the agency view that our concept of cause derives from our experiences as agents. Huw Price (1996) used that view to argue that the direction of causation is a projection of the direction of the means-ends relation. I, on the other hand, am a realist about powers and natures, so I believe that if it is not in the nature of a thing to do X, it cannot be manipulated to achieve X. This may create a tension between the capacity approach and Price's perspectivalism, but the discussion of this issue is better left for another occasion.

References

Aristotle (1928) *Metaphysica*. Translated by W. D. Ross. Oxford: Clarendon Press.

Cartwright, N. (1989) *Nature's Capacities and Their Measurement*. Oxford: Clarendon Press.

Cartwright, N. (1999) *The Dappled World*. Cambridge: Cambridge University Press.

Cartwright, N. (This volume) Modularity: It can—and generally does—fail. In D. Costantini, M. C. Galavotti, and P. Suppes, *Stochastic Causality*. Stanford: CSLI Publications.

Davidson, D. (1967) Causal relations. *The Journal of Philosophy*, 64, 691-703.

Davidson, D. (1968) Actions, reasons, and causes. In A. White (Ed.), *The Philosophy of Action*. Oxford: Oxford University Press, 79-94.

Dretske, F. (1969) *Seeing and Knowing*. London: Routledge & Kegan Paul.

Gopnik, A. and Meltzoff, A. (1997) *Words, Thoughts, and Theories*. Cambridge: The MIT Press.

Gopnik, A., Glymour, C. and Sobel, D. (forthcoming). Causal maps: A cognitive and computational account of theory formation. In P. Gardenfors, K. Kijania-Placek, and J. Wolenski (Eds.) *Logic, Methdology and Philosophy of Science XI*. Dordrecht: Kluwer Academic Publishers.

Harre, R. and Madden, E. (1975) *Causal Powers*. Oxford: Oxford University Press.

Hume, D. (1739/1962) *A Treatise of Human Nature*. Glasgow: Fontana.

Irzik, G. (1996). Can causes be reduced to correlations? *British Journal of Philosophy* 36, 273-289.

Leslie, A. (1982) The perception of causality in infants. *Perception*, 11, 173-186.

Leslie, A. (1984) Spatiotemporal continuity and the perception of causality in infants. *Perception*, 13, 287-305.

Leslie, A. and Keeble, S. (1987) Do six-month-old infants perceive causality? *Cognition*, 25, 265-288.

Mackie, J. L. (1980) *The Cement of the Universe*. Oxford: Clarendon Press.

Michotte, A. (1963) *The Perception of Causality*. London: Methuen.

Papineau, D. (1985) Causal asymmetry. *British Journal of Philosophy,* 36, 273-289.

Papineau, D. (1991) Correlations and causes. *British Journal of Philosophy,* 42, 397-412.

Papineau, D. (1993) Can we reduce causal direction to probabilities? In D. Hull, M. Forbes and K. Okruhlik (Eds.), *PSA 1992,* Vol. 2, East Lansing: Philosophy of Science Association, 238-252.

Pearl, J. and Verma, T. (1994). A theory of inferred causation. In D. Prawitz, B. Skyrms, and D. Westerstahl (Eds.), *Logic, Methodology, and Philosophy of Science IX*. Amsterdam: Elsevier, 789-811.

Price, H. (1996) *Time's Arrow and Archimedes' Point*. New York: Oxford University Press.

Spirtes, P., Glymour, C. and Scheines, R. (1993) *Causation, Prediction and Search*. New York: Springer-Verlag.

Spohn, W. (This volume) Bayesian nets are all there is to causal dependence. In D. Costantini, M. C. Galavotti, and P. Suppes, *Stochastic Causality.*

Suppes, P. (1970) *A Probabilistic Theory of Causality*. Amsterdam: North Holland.

Woodward, J. (1990) Supervenience and singular causal statements. In D. Knowles (Ed.), *Explanation and its Limits*. Cambridge: Cambridge University Press, 211-246.

6

Causation in the Special Sciences: The Case for Pragmatism

HUW PRICE

University of Sydney

1 Introduction

One of the jobs of philosophers of the special sciences is to connect the local concerns of particular disciplines with those of philosophy in general. The two-way complexities of this task are well illustrated by the case of causation. On the one hand—from the outside, as it were—philosophers interested in general issues about causation are prone to turn to the special sciences for real-life examples of the use of causal notions. On the other hand, from the inside, the special disciplines themselves throw up philosophical puzzles in which the notion of causation plays a role. When does correlation indicate causation, for example? Physics and economics both generate hard cases of this kind.

In principle, then, a philosopher of a discipline such as physics or economics occupies a rather exposed position, liable to be called as expert witness about causation from inside and out—by people who know more about the special discipline, and by people who know more about relevant parts of philosophy at large. This dual role may seem a trifle self-contradictory. After all, to address a causal puzzle within physics (say) a philosopher of physics needs to hold fixed a general philosophical account of causation; and therefore hasn't the luxury of remaining uncommitted about the philosophy of causation in general, the kind of cautious player who awaits the verdict of

Stochastic Causality.
Maria Carla Galavotti, Patrick Suppes and Domenico Costantini (eds.).
Copyright © 2001, CSLI Publications.

physics on such matters. But progress is surely possible by a kind of reflective equilibrium, an informed interplay between the two kinds of constraints, philosophical and scientific. The difficult task, the task which falls especially to the philosopher of the special disciplines, is to open and keep open the dialogue.

This paper is an attempt at that task. In particular, I want to recommend a particular way of thinking about causation—a kind of pragmatism—to philosophers of sciences such as physics and economics. The term pragmatism means several things in contemporary philosophy, and so I'll say more in a moment about the sense I have in mind. In the sense in question, pragmatism about causation is an uncommon view—and particularly so, I think, among philosophers of the special sciences. But it deserves to be taken much more seriously, in my view, especially within the philosophy of physics and economics. For pragmatism about causation throws important light on puzzles within these disciplines, and considerations stemming from these disciplines provide some of the main reasons for endorsing the pragmatic approach.

I suspect that some of the resistance to this pragmatic approach within the philosophy of the special sciences stems from the feeling that is it is too 'subjective', not sufficiently 'realist', for the purposes of the sciences concerned. Philosophers of the special sciences often show a commendable concern to defend the 'seriousness' of the sciences on which their own discipline is parasitic. However, the usual maps of the relevant philosophical territory are sometimes rather crude, I think. There are several varieties of realism, and several varieties of subjectivism, not often well distinguished. One of my aims here is philosophical cartography: I want to draw some maps to clarify on the one hand, what it means to be a realist about causation, in the philosophy of physics or economics; and on the other, the sense in which the kind of pragmatism I have in mind is and is not a subjectivist view. Roughly, I hope to show both that the view is not subjectivist in the sense most inimical to realism, and that to the extent that realism is an attractive position about causation, it is compatible with the kind of pragmatism I have in mind.

I also want to call attention to a fallacy, as I see it, in a popular defence of realism about notions such as causation. The argument in question relies on a supposed analogy between causation and the theoretical postulates of the special sciences. The suggestion is that postulating a causal relation is 'just like' postulating electrons or any other theoretical entity, so that (among other things) realism about causation is no more problematic than—or at least no different in kind from—realism about electrons. In my view, this supposed analogy is seriously flawed, for reasons which turn out to support a pragmatic approach to causation.

2   Three Kinds of Pragmatism

What is a pragmatic account of causation, or of any other philosophical topic? The term pragmatism is associated with at least three distinct philosophical views. One view, stemming from Peirce, takes pragmatism to be primarily a doctrine about truth: the view that truth is to be identified with some epistemological notion such as warranted assertibility. Another view, more closely associated with James and Dewey, takes the key to pragmatism to be an appeal to a notion of *success*. Like Peirce's pragmatism, this second view can issue in an account of truth. For example, it might be said that true beliefs are those which lead to *successful* behaviour. But the approach itself is not tied to such an account of truth: a pragmatist of this second sort might equally give a success-based account of other notions, such as mental content.

A third form of pragmatism, and the one which interests me here, is the view that a philosophical account of a problematic notion—that of causation itself, for example—needs to begin by playing close attention to the role of the concept concerned in the *practice* of the creatures who use it. Indeed, the need to explain the use of a notion in the lives of ordinary speakers is often the original motivation for an account of this kind. Causal notions and their kin are ubiquitous in the everyday talk of ordinary people. Pragmatists argue that we cannot hope to explain this anthropological fact if we begin where metaphysics traditionally begins, at the level of the objects themselves—if we ask what causation *is,* if we begin by looking for something for causation to *be,* which will explain all these uses. Instead, pragmatists think, we need to start with the practice of *using* such notions, and to ask what role such notions play in the lives of the creatures concerned—why creatures like us should have come to describe the world in these causal terms.

The pragmatic approach to causality and related notions is certainly not new. Ramsey's views on probability, and his late views on causation and physical necessity, are of this kind, for example (1978a, 1978b). Take probability, for instance. Ramsey's view is that an account of probability needs to begin with the role of probability judgments in decision making. This connection needs to be there from the very beginning. We could not give a satisfactory account of probability which left it out initially, and then proceeded to note that the notion thus characterised had some relevance to decision contexts.

However, I think the distinctive character of the pragmatic approach is not well understood in contemporary philosophy. This is partly a result of some confusing terminology left to us by tradition. Pragmatism of this kind is thought to be *subjectivist,* or *non-objectivist.* But these terms too apply to several distinct approaches, often poorly distinguished, and pragmatism is easily mischaracterised. This kind of confusion contributes to misunderstanding of the relationship between pragmatism and realism, in my view. Roughly, realism requires objectivism in one sense, while pragmatism of this kind denies objectivism in another sense. The two views are thus less in conflict than commonly assumed.

In fact, in turns out to be useful to distinguish *three* kinds of objectivity—three somewhat independent respects in which a view may vary between objectivism and subjectivism. In one of these respects, subjectivism is indeed in conflict with realism, as intuitively understood. As we shall see, however, a philosophical view may be objectivist in this realism-related respect and yet subjectivist in either of the other two respects. In this way, pragmatism about causation remains compatible with realism: The respect in which pragmatism is subjectivist is not the respect in which realism requires objectivism, and the two views are not in conflict.

3   Three Kinds of Objectivity

What are the three varieties of subjectivity and objectivity? For present purposes, applied to the case of causation, they may be characterised as follows:

1.   Causation is *ontologically subjective* if the *existence* of causal relations depends on the presence of minds, speakers, observers or the like; if causal states of affairs are in this ontological sense mind- or observer-dependent.

2.   Causation is *topic-subjective* if talk of causation is in part talk *about* speakers, agents, or humans—for example, if all causal claims are in part about our own psychological states.

3.   Causation is *practice-subjective* if an adequate philosophical account of causation needs to make central reference to the role of the concept in the lives and practice of creatures who use it.

Some cautionary notes. Obviously, these characterisations are somewhat rough. In particular, I have not been careful about marking the fact that some of these characterisations apply more naturally to causation itself—the object or 'thing' in the world—and some to the term or concept. (More on this later.) And the list may not be exhaustive, in the sense that there may be other interesting varieties of subjectivity and objectivity which could usefully be added. Still, this three-way classification provides some useful distinctions. In particular, it will enable me to maintain that concerns about realism relate mainly to the ontological axis, whereas the pragmatism I advocate about causation involves practice subjectivity.

More on the last point in a moment. First, let me emphasize that pragmatism about causation is not the view that when we talk of causation we are talking *about* ourselves, in whole or in part. The latter view would involve topic-subjectivity, not practice-subjectivity. The corresponding distinction is often missed, I think, in the analogous case of subjectivism about probability. As a result, the view I attributed to Ramsey—the interesting and in my view plausible claim that probability cannot be characterised without reference to its role in decision-making—is confused with the view that speakers use probabilistic terms to talk *about* their own mental states.

If the difference between these views isn't clear, think of what the two views say about a claim such as, "It will probably rain tomorrow." According to the topic-subjective account, someone who says this is talking about her own state of mind—she is saying, in effect, that she is *confident* that it will rain tomorrow. All probability claims are thus psychological in content, at least in part. Not so according to the practice-subjective account: probability is not a psychological concept, on this view. True, an adequate philosophical account of the concept will need to mention psychological notions, in virtue of mentioning the role of the concept in decision making. But to say that an explication of the function of a concept needs to mention psychological notions is not to say that that concept has a psychological content: of course not, for otherwise all human language would have a psychological content, simply in virtue of the fact that an account of the function of language needs to be grounded in psychology. (To put it more simply, we might explain the function of the utterance "It will rain" by saying that it is used to give voice to the *belief* that it will rain; but this does not imply that the utterance is *about* that belief.)

I want to say a corresponding thing about causation. In my view, an adequate philosophical account of causation needs to begin with its role in the lives of *agents*, creatures who have the primitive experience of intervening in the world in pursuit of their ends. I have defended such a view in a number of places (see Price 1991, 1992a, 1992b, Menzies & Price 1993, and especially Price 1996, ch. 6), and noted its affinity to a late view of Ramsey (1978b). Here, I simply want to emphasize that the view is not the topic-subjective claim that talk of causation is talk *about* agents or agency, but rather the practice-subjective doctrine that we don't understand the notion of causation—as philosophers, as it were—until we understand its origins in the lives and practice of agents such as ourselves.

So much for the distinction between topic-subjectivism and practice-subjectivism. It remains to point out that neither commits us to ontological subjectivism, in any interesting sense. Because topic-subjectivism treats causation (or whatever) as in part a matter of psychology, it is ontologically subjective in the boring sense in which psychology itself is: if there were no thinking creatures, there would be no psychological states. As for practice-subjectivism, it is not even ontologically subjective in this uninteresting way. If the concept of causation is essentially tied to our experience as agents, as my kind of practice-subjective pragmatism maintains, then of course the concept would not arise in a world without agents. But this does not make it appropriate to say that if there had been no agents there would have been no causation. Pragmatism does not conflict with realism in that sense. In other words, as I noted above, realism seems to require ontological objectivism, and my kind of pragmatism about causation requires practice-subjectivism. These two requirements are not in conflict.

4   Realism Responds?

At this point, some realists will feel that the compromise has been achieved at the cost—from my point of view—of making pragmatism a rather uninteresting view. These realists have no objection to the view that an account of the *concept* of causation is practice-subjective, but see the compromise as resting on the fact that this is quite compatible with a demand for realist account of causation itself—that is, of the 'thing in the world', rather than simply the concept by means of which we refer to it.

I agree that by the lights of such a realist account, if there could be such a thing, the pragmatist view would be at best a supporting act, an interesting prelude to the main game. However, I want to cast doubt on the claim that there could be a main attraction of the kind the realist promises. I think there are good reasons for doubting such claims in philosophy in general, but here my goal is more specific: I want to show that there are particular reasons for being suspicious of this promise in the case of causation—for thinking that pragmatism retains centre stage in this case, whatever its fate elsewhere.

My argument turns on two main ingredients: first, a distinction between two species or grades of realism, one stronger than the other, and second, a rejection of an often-claimed analogy between causation and the theoretical entities of first-order science. Roughly, I want to argue that the realist move just canvassed requires the stronger of the two grades of realism, but that this is even more problematic in the case of causation than it is in the supposedly analogous case of the theoretical entities of first-order science. In the sense in which realism about causation is defensible, then, it is also compatible with the view that pragmatism is the central ingredient in a philosophical account of causation.

In view of the importance of the supposed analogy, it will be useful to characterise the two grades of realism initially as two different responses to antirealism about theoretical entities. And it will be instructive, I think, to do this against a brief sketch of the most relevant historical background. I include this sketch for two reasons. First, I commented earlier on a feeling in contemporary philosophy of science that antirealism shows a lack of due respect for science, a lack of seriousness about the special disciplines. Sometimes antirealism seems to be thought of as the product of a recent, aberrant, anti-scientific turn. Yet a little history makes clear that the very opposite is true: for three hundred years, antirealism has often seemed the proper consequence of the kind of deep respect for science embodied in the empiricist tradition—of the desire, especially, to rid science of the vestiges of a degenerate and unempirical metaphysical tradition. This impression may have been wrong, of course, but contemporary friends of science—philosophers of physics and economics, for example—do well to be careful. Prescientific metaphysics may lurk in the guise of scientific realism.

The second virtue of the following sketch lies in the fact that another feature of this historical tradition, a feature present from its beginnings in the seventeenth century, is the desire to distinguish what is 'objective' in the world from what in some sense 'comes from us'. I want to take the opportunity to link some of the milestones of this long project to the three-way classification of kinds of subjectivity set out above, and hence to the kind of pragmatism I advocate about causation.

5 The Origins of the Debate

The issue of objectivity is a major concern of many of the great natural philosophers of the seventeenth century, writers such as Descartes, Galileo, Boyle and Locke. One issue which concerns these thinkers is the distinction between what Boyle and Locke call the primary and secondary qualities. Roughly, the primary qualities are those which exist and have their natures independently of sentient observers, and are therefore apt to be studied by a science of the natural world. The secondary qualities are those of which this is not true—those which are 'observer-dependent', in some sense. Here, for example, is a beautiful early account of the distinction, from Galileo:

> I feel myself impelled by the necessity, as soon as I conceive a piece of matter or corporeal substance, of conceiving that in its own nature it is bounded and figured in such and such a figure, that in relation to others it is large or small, that it is in this or that place, in this or that time, that it is in motion or remains at rest, that it touches or does not touch another body, that it is single, few, or many; in short by no imagination can a body be separated from such conditions; but that it must be white or red, bitter or sweet, sounding or mute, of a pleasant or unpleasant odour, I do not perceive my mind forced to

acknowledge it necessarily accompanied by such conditions; so if the senses were not the escorts, perhaps the reason or the imagination by itself would never have arrived at them. Hence I think that these tastes, odours, colours, etc., on the side of the object in which they seem to exist, are nothing else than mere names, but hold their residence solely in the sensitive body; so that if the animal were removed, every such quality would be abolished and annihilated.

Il Saggiatore (quoted by Burtt 1932, p. 75.)

In this passage, the distinction is drawn in terms of the 'location' of the two kinds of qualities. Primary qualities lie in the world, secondary qualities (despite appearances and common beliefs to the contrary) lie in us. Thus secondary qualities are seen as topic-subjective, in the terms distinguished above. As a result, they are also ontologically subjective: their existence depends on the presence of the 'sensitive body'.

In the tradition which follows, however, this ontological issue soon becomes blurred. Perhaps secondary qualities are in the world, after all, but are of a subject-involving nature—in ascribing such a property to the world we say that the world is such as to affect sentient creatures in certain ways. This is closer to Locke's view, and while it remains topic-subjective, but is less ontologically subjective, for presumably the world might have such dispositional properties in the absence of any *actual* sentient creatures.

A Lockean account of this kind is easily transformed into a view which is practice-subjective, rather than topic-subjective. For example, suppose that the Lockean view says that to be red is to be disposed to produce a certain response R in humans. Then there is a corresponding practice-subjective account of the concept red, a central feature of which will relate the *use* of this concept to a speaker's having response R. (On the most crude account, the concept is simply used to give voice to this response.) Thus as in the case of subjectivism about probability, the topic-subjective view interprets "That's red" as in part *about* human psychology, whereas the practice-subjective view does not. (I have argued elsewhere (Price 1998) that this practice-subjective approach to pragmatism has general advantages compared to its Lockean twin. Here, the important point is just that there are two distinct possibilities.)

However the subjectivism of the secondary qualities is characterised, the issue arises as to whether *any* properties are purely primary, in the original sense. Perhaps the properties in terms of which we describe the world are never the essential properties of the things in themselves; perhaps we only ever know and describe things in terms of their appearances from our own point of view. This thought crops up in later philosophy of science in 'structuralist' or 'dispositionalist' views of theories, which claim that science always describes only the general patterns or structures in which reality manifests itself to us, not reality itself. One source of this view is Kant, whom Putnam (1981, p. 60) suggests may be read as arguing that all properties are secondary. (Putnam calls the result a kind of pragmatism.)

These structuralist views illustrate a very deep tension in empiricism since the seventeenth century, between the metaphysical desire to know the world as it really is, via empirical science, on the one hand; and the epistemological desire that our knowledge claims be well-grounded in observation, on the other. Metaphysics pulls us outwards, epistemology pulls us inwards. In structuralist views, the view that science searches for inner natures has been dropped, in favour of the view that it searches for something more accessible, by empiricist lights. This is far from an isolated case. In empiricism in general, the advantage of relatively subjective approaches is thought to be that they avoid unjustified inferences to things 'beyond' the observable.

From Hume onwards, moreover, the inward pull of epistemology has been reinforced by a concern about meaning. If, as many empiricists have held, the acquisition of meaning needs ultimately to be grounded in a novice speaker's observational experiences, then meaningful talk about unobservable matters seems impossible. Metaphysical claims seem in danger of being empty, rather than merely unjustified.

These considerations are the main source of the problem of the status of theoretical terms in science, and the motivation for antirealism about such terms. As I noted earlier, these empiricist antirealists are not enemies of science, as a casual reader of contemporary debates might assume. On the contrary, the antirealists' goal is to save the truly scientific core of what passes for science, by ridding it of the remnants of the old metaphysics—to free their beloved *empirical* science from the grip of a degenerate and anti-empiricist style of philosophy.

One of the more interesting forms of antirealism is instrumentalism, which would save science from metaphysics by denying that the function of theoretical discourse is to describe a reality 'beyond' observation (and so denies that theoretical terms have a referential function). On the contrary, an instrumentalist says, the significance of theoretical terms lies in their role in a particular human practice—a practice whose function lies in its ability to predict future observations. So this approach is pragmatic, in the third sense I distinguished at the beginning.

Instrumentalism thus illustrates one way in which pragmatism may become a global position, or at least a position with very wide application. I mentioned another view of this kind a moment ago, in Putnam's interpretation of Kant, as a philosopher who believes that all properties are secondary properties. It may seem that these views pose a problem for the project of defending a pragmatic approach to causation, in the sense that they promise an embarrassment of riches. If pragmatism promises to become a global view, what need is there for argument which picks out the case of causation?

However, there is no substantial tension here with my present project. My present argument turns on the fact that realism and objectivism about causation are importantly different from, and more problematic than, the corresponding views about the theoretical postulates of science. Put another way, I am arguing that there are reasons for pragmatism about talk of causation which are not reasons for pragmatism about (say) talk of electrons. The differences on which this argument depends are compatible with the possibility that there might be other reasons for endorsing pragmatism of some global variety.

6 Two Grades of Realism

There are two popular ways of trying to avoid antirealism about the theoretical entities of science. One, which I shall call *strong realism,* relies on so-called 'inference to best explanation'. On this view, our reason for believing in the reality of (at least some of) the theoretical entities and states of affairs postulated by science is that if these things exist or obtain, they provided the best causal explanation of the available observational data, broadly construed. This approach thus rejects Russell's (1917) view that science should avoid inferred entities wherever possible, and defends the legitimacy of such inferences, at least in causal–explanatory contexts. The causal–explanatory project provides the framework which legitimates what many empiricists saw as idle and ill-founded metaphysics.

The second popular brand of realism in contemporary philosophy of science is what may be called *weak* or *minimal* realism. This view simply takes the existence claims of science at face value, and rejects any 'additional' metaphysical or philosophical viewpoint from which it would make sense to ask 'Do these things (electrons, for example) *really* exist?' The key to weak realism is the rejection of a standpoint for ontology beyond that of science. The most famous proponent of this view is therefore perhaps Quine, for whom the rejection of such an 'external' ontological standpoint is the key doctrine of papers such as 'On What There Is' (Quine 1953, ch. 1). More recently, for example, it is prominently embodied in Arthur Fine's (1986, ch. 7) advocacy of what he terms the 'natural ontological attitude'.

I am following convention in calling this view a species of realism. However, it is also instructive to see the view as rejecting the traditional realist–antirealist debate altogether, at least as that debate arises within the empiricist tradition. Roughly, a minimalist wants to say that within this debate, both sides presuppose that there is a legitimate standpoint for philosophy, apart from that of science, from which questions such as the following make sense: Do the theoretical entities postulated by our best scientific theories *really* exist? Do the theoretical terms of these theories *really* refer, and if so to what? Do theoretical claims *really* have truth values? Realists and antirealists give different answers to these questions, but agree that the questions themselves (or variants of them) are well-posed, in the intended 'supra-scientific' sense. But minimalism itself challenges this assumption, and thus rejects the debate. As Fine (1986, ch. 8) makes clear, the view is neither realism nor antirealism, *in the old sense*.

Among the challenges that weak realism offers to its stronger cousin is a dilemma concerning inference to best explanation itself. Is the employment of this inference itself part of science? If so, then it doesn't take us beyond what the weak realist already accepts—it doesn't yield an ontological standpoint beyond science. (An instrumentalist of the above sort should have no objection to it, for example.) If not, then how is it to be justified? How could it have a justification which wasn't already available to science?

My sympathies lie with the weak realist on this point. However, my present purpose is not to adjudicate on the issue of the proper form of realism about theoretical postulates, but to use that issue as background to the issue concerning realism about causation. In particular, I want to argue that the case against strong realism is even stronger in the case of causation than in the case of theoretical entities such as electrons. Someone who wants to argue for strong realism about causation by analogy with theoretical entities thus faces two kinds of challenge: those intrinsic to the debate in the philosophy of science in general, and those which depend on the fact that the analogy itself is problematic, in the way I want to explain.

7 Is Causation a Theoretical Postulate?

As I have noted, some philosophers suggest that causation and other modal notions are on a par with the theoretical entities postulated by science. (If there is a difference, it is simply that unlike electrons, causation is already a part of prescientific 'folk theory'.) On this view, our reasons for treating causation realistically are essentially the same as those for treating electrons realistically.

If what is at issue is strong realism about causation, however, then we run into problems. As we have seen, the main argument for strong realism about theoretical entities goes in terms of inference to explanatory *causes*. But this reason simply takes the notion of causation for granted, and therefore can't be applied in *support* of realism about causation. In this context, the supposed role of inference to best explanation is epistemological—it is supposed to *justify* a belief in the reality of entities of a certain kind. My point is that such an attempt at justification would be viciously circular in the case of causation itself, in virtue of the fact by the realist's own lights, the inference presupposes realism about explanatory causes.

A similar point can also be made in a metaphysical key. Suppose we are interested in investigating the nature of causation, in the manner of the realist we imagined in Section 4, who saw pragmatism as a conceptual prelude to metaphysics. Again, the analogy with the theoretical entities of science breaks down. In science we learn more about an entity by learning more about its causes and effects; we postulate theoretical entities in terms of causes and effects. These techniques reduce to nonsense if we try to apply them to investigate causation itself. Causation is the medium in which the investigation is conducted, the thread which leads to unseen objects. Treat the thread itself as unknown, and we are simply blind.

To make the same point in philosophical vocabulary which will be familiar to many, consider David Lewis's account of theoretical reduction (1970, 1972). According to Lewis, electrons are whatever entities actually play the 'causal roles' defined by electron theory. To accept electron theory is to accept that there is something in reality which plays these roles. What we know about what electrons actually are is what we find out by empirical means about the entities which do these causal jobs. Once again, this approach would be circular if we were to try to apply it to causation itself. The term 'causation' is employed in the reductive machinery itself, and therefore cannot be processed by that machinery.

It is true that the immediate problem can be avoided if causal roles are replaced by semantic roles, so that, for example, electrons are said to be the *referents* of the appropriate terms in electron theory. But in this case the circularity re-emerges if the approach is applied to the semantic notions themselves—or if reference is defined in causal terms!

The immediate problem can also be avoided if causation is analysed in terms of other modal notions—if causation is reduced to counterfactuals and possible worlds, for example. However, it then re-emerges if we attempt to use the same argument in favour of realism about those notions themselves. To avoid this complexity, I shall assume for the moment that our topic is not causation but modality in general.

The general problem for modal realists may then be characterised as follows, in two keys. In the epistemological key, it is that if the justification of theoretical beliefs in science relies on modal presuppositions, such as those that underlie inference to best explanation, then the justification of modal beliefs cannot be of the same kind, on pain of a threat of vicious circularity—the threat that the proposed justification *presupposes* some of the modal claims in question. In the metaphysical key, it is that the *investigation* of modal reality cannot rely on the standard tools of empirical science, to the extent that these tools are themselves constructed of modal materials.

Thus the supposed analogy between causation (and modal notions generally), on the one hand, and the theoretical entities of first-order science, on the other, breaks down in important ways. Even if strong realism were unproblematic with respect to the theoretical postulates of science, that would not count in favour of strong realism about causation. Indeed, the conclusion seems even worse than that, from the strong realist's point of view. The argument suggests that strong realism about theoretical entities is bound to be in trouble, by the realist's own lights, unless strong realism about causation is already secure. And the point applies whether the realist's concerns are epistemological or metaphysical: whether she wants to justify a belief in the reality of causal relations, or whether she wants to investigate their nature.

So much for the bad news for strong realism about causation. Now to the good news for pragmatism. In the epistemological key, the point just made is that the inference to modal conclusions from non-modal premises does not parallel that to theoretical beliefs in science. Inferences of the latter kind presuppose the relevant modal background in a way in which inferences of the former kind could not. Let us call this the problem of *upward* inference (inference *to* modal conclusions).

Modal realists also face a problem about inferences in the opposite direction. We may call it the problem of *downward* inference, or the *application problem*. In turns on the issue as to why objective modal facts should 'matter' to us in the way that they do. Why should we take them into account in decision-making, for example? Consider, for example, an account of possibility in terms of the existence of real 'possible worlds', distinct from our own world. If there were such worlds, why should they matter to us when we make decisions in this world? Wouldn't everything be just the same *for us* even if those other worlds didn't exist? Similar problems arise for other objectivist views about modal notions. If probability is objective, why should it have its assumed relevance to human decision making?

As in the case of the problem of upward inference, this downward or application problem doesn't arise with respect to theoretical entities in science. In that case, roughly speaking, application or relevance is simply *defined* in terms of causality. The relevance of theoretical commitments lies in their causal consequences. But again, this would be viciously circular if we tried to apply it to causation itself.

In my view, the pro-pragmatist lesson to be drawn from these considerations is that in considering causation, probability, and other modal notions, philosophy needs to begin with the practice of which the upward and downward inferences are already a part. If we think of the issue of the status of this practice as an adjunct to a strongly realist account, an account which begins by postulating the existence of a realm of entities of a certain kind, then it will always seem a mystery why talk of such entities should play the role that it does in our lives (why probability judgements should have the role they do in decision, for example).

In a sense, the supposed analogy with theoretical entities simply highlights this difficulty. It encourages us to think of causation and the like as part of the furniture of the world, on a par with the aspects of the world investigated by the sciences. And this very parity then becomes the problem. Why should particular pieces of furniture—the modal pieces, not the others—have the special significance that they do? (Don't say that we can ask the same question about electrons. In that case the answer is in causal terms.)

To sum up. The supposed analogy with the theoretical entities of science fails as a defence of 'strong' realism about causality and other modal notions. Moreover, because the failure turns on the role that the modal notions themselves play in arguments for strong scientific realism, their failure exposes a deep problem for modal realism, and a deep advantage of the pragmatist approach.

What is the realist about causation to do at this point? One option is to be less ambitious, by espousing only weak or minimal realism. Won't the analogy with theoretical entities survive, so long as we are weak realists in both cases?

In my view, its survival is doubtful, even in this context. If inference to best causal explanation remains an important tool within science, it remains a tool which cannot be used by a scientist to lay bare the causal relation itself. So the analogy still breaks down. However, for present purposes I want to emphasize a different point concerning weak realism about causation and modality, namely that this form of realism is perfectly compatible with pragmatism, in the third sense outlined at the beginning.

8   Pragmatism and Minimal Realism

The defining doctrine of weak or minimal realism is that the first-order claims of science or ordinary usage don't need to be 'validated' from some external philosophical perspective. This view is sometimes characterised as a kind of quietism. It denies the need for—and, usually, the coherency of—a certain kind of philosophical enquiry. Clearly, this kind of quietism is as intelligible with respect to our commitments about causation as it is with respect to those about electrons. In both cases it amounts to the same advice: Stop doing what tradition has done. Walk away from the realist–antirealist debate, at least in this case.

Sometimes quietism of this kind goes as far as recommending that we walk away from philosophy, or at least from a major part of its concerns. However, this stronger recommendation is inessential, and in my view misguided. To reject an externalist ontological viewpoint need not be to reject all kinds of philosophical reflection on the topics in question. In particular, it is quite compatible with taking seriously questions like these: Why do we humans go in for talk of causation and other modal notions? What role does this 'talk' play in our lives, and to what extent does this role depend on particular features of our situation (on the fact that we are agents, for example)? How, if at all, does causal and modal talk differ in these respects from talk of electrons, or tables, or numbers, or whatever?

These are precisely the questions that a pragmatist seeks to answer—a pragmatist in my third sense, the sense I take to be exemplified by Ramsey's account of probability, and especially by his late account of causality and laws of nature. In my view, this kind of pragmatism embodies what turn out to be the philosophically interesting issues about causation and the modal notions generally, once we set aside strong realism, and the ill-grounded idea that the status of modal notions is 'just like' that of the theoretical postulates of the special sciences.

One aspect of the realism thus rejected is the view that in order to validate talk of causation in science and ordinary life, we need to find something in the physical world with which to 'identify' causation. I have argued that this reductive project cannot simply ride on the back of scientific reductionism, for the latter takes causal notions for granted in a way in which the former cannot. However, the project has many admirers in contemporary philosophy, and in rejecting it in the case of causation we are swimming against a strong tide. So I want to close by mentioning another kind of objection to such reductive accounts of causation, which I have developed in much more detail elsewhere (see especially Price 1996, ch. 6). This objection turns on the claim that at least one central feature of causation—roughly, its 'directedness'—simply isn't there to be 'found' in the appropriate place in physical reality. Again, pragmatism does well where realism fails. The feature in question falls naturally into place in an account of the human origins of the concept of causation.

9 Realism, Pragmatism and the Arrow of Causation

Causation seems 'directed' in two senses. First, the causal relation is asymmetric, in the sense that if A is a cause of B then B is not a cause of A. Second, this 'causal arrow' is strongly aligned past-to-future. Effects don't occur *before* their causes. As a result, causation comprises one of the main ways in which the world seems asymmetric in time. So there is a puzzle here: Where does this asymmetry come from, especially in the light of the temporal symmetry of the underlying physical laws. But this puzzle is secondary to that of the causal arrow itself: What distinguishes cause and effect? What determines which way the causal arrow 'points'?

Let us think about these puzzles from the point of view of a strong realist, who thinks that there is a well-posed issue about the real nature of causation, of the 'thing in the world'. From such a point of view, an account of the causal arrow seems to require two things: an understanding of the asymmetry, or 'directedness' of the real relation itself; and an answer to the question as to why this intrinsic directionality should show the temporal preference that it does.

Most contemporary realists would be reluctant to treat these aspects of causation as brute facts about the world. (One reason for reluctance is the difficulty of connecting brute facts with practice. How could we tell if the brute facts were otherwise, for example?) Many, therefore, would prefer to seek some reductive account of causation, and some appropriate connection with other temporal aspects of the world.

The approach then becomes what I have called the *third arrow strategy* (Price 1996, ch. 6). This strategy tries to account for the directedness of causation by reducing causation to some feature of the physical world which is itself asymmetric or directed in the appropriate way—something which both has its own internal 'arrow' to give us the distinction between cause and effect, and something such that this internal arrow turns out to have the temporal alignment we find in the case of causation. This 'third arrow' then links those of causation and time.

Unfortunately for strong realism, the third arrow strategy is hard to reconcile with contemporary physics. Modern physics doesn't seem to provide a third arrow, of the appropriate kind. The favoured candidate is the time-asymmetry associated with the second law of thermodynamics, but this turns out to be inadequate for the task at hand (see Price 1992a, 1996). For example, it doesn't work at a sub-statistical level, to explain our causal intuitions about microphysics.

Realists have a fall-back position. They can concede that the causal relation is not really 'directed', in any non-conventional sense. The relation remains asymmetric in the logical sense—if A is a cause of B then B is not a cause of A—but the distinction between cause and effect is now regarded as merely conventional. In other words, it is analogous to the conventional distinction between up trains and down trains (on a railway line with no significant overall gradient). So long as the labeling convention imports a temporal asymmetry, by fiat or in fact, it saves the phenomena. Cause and effect are distinguished, and aligned in time.

This conventionalist strategy plays into the pragmatists' hands, however. The strategy comes in three main versions:

1. Follow Hume in stipulating that causes are the *earlier* and effects the *later* of pairs of events related in some otherwise time-symmetric way (e.g., as in Hume, by a relation of constant conjunction).

2. Follow Reichenbach (1956), so that the stipulation refers not to earlier and later, but to orientation with respect to some physical temporal 'signpost', such as the thermodynamic asymmetry.

3. Take the relevant 'signpost' to be provided by our perspective as agents: say that the cause is the element of the pair which we could 'manipulate', in principle, if the other element were our desired end. In this case the temporal orientation follows from our temporal orientation as agents.

The three options come to much the same thing in the end, I think, but because (3) promises to explain why such a labeling convention should be useful to us, it gets the genealogy right on the first pass. (1) and (2) need an additional step, to connect the convention with our own practice.

One way to see this is to ask why the causal arrow matters, on this view. Why should we bother to treat causation as temporally asymmetric? In other words, why should temporal information be embedded in the labeling convention for causation in this way? Option (3) contains the beginnings of an answer to this question, because it connects the asymmetric labeling convention to something which matters to us, and something which has connections with time. As agents, we act for the future, on the basis of information about the past.

However, the attractiveness of (3) is just the attractiveness of pragmatism. In backing-off as far as this, in other words, a strong realist approach has conceded that strong or reductive realism has nothing distinctive to tell us about one of the most distinctive features of causation, its the directedness and temporal orientation. Pragmatism thus has an important advantage over strong forms of realism in this case, an advantage which is quite independent of general issues between realism and its critics.

10 Conclusion: The Case for Pragmatism

I remarked earlier that within the philosophy of the special sciences, seriousness about science is sometimes taken to require realism about causation, of a kind which conflicts with pragmatism. I have tried to counter this thought, in a number of ways. In particular, I have argued that pragmatism is compatible with the kind of realism which perhaps takes science most seriously, in not taking it to be subject to 'validation' from some distinctively philosophical vantage point. I have also claimed that the breakdown of supposed analogy between causation and theoretical entities not only makes any stronger realism more problematic for causation than it is in science itself, but also highlights the central role of pragmatic considerations in any adequate account of causation.

If there is a single conclusion to be emphasized, it is this one. Being realist about causation is *not* like being realist about electrons, or any of the other postulates of the special sciences. Causation is different, and it is pragmatism, not realism, which offers an account of the difference.

References

Burtt, E. A. (1932) The Metaphysical Foundations of Modern Physical Science. London: Routledge & Kegan Paul.

Fine, A. (1986) The Shaky Game. Chicago: University of Chicago Press.

Lewis, D. (1970) How to define theoretical terms. Journal of Philosophy, 67, 427–446.

Lewis, D. (1972) Psychophysical and theoretical identifications. Australasian Journal of Philosophy, 50, 249–258.

Menzies P. and Price, H. (1993) Causation as a secondary quality. British Journal for the Philosophy of Science, 44, 187–203.

Price, H. (1991) Agency and probabilistic causality. British Journal for the Philosophy of Science, 42, 157–176.

Price, H. (1992a) Agency and causal asymmetry. Mind, 101, 501–520.

Price, H. (1992b) The direction of causation: Ramsey's ultimate contingency. In D. Hull, M. Forbes and K. Okruhlik (Eds.), PSA 1992, Vol. 2. Philosophy of Science Association, 253–267.

Price, H. (1996) Time's Arrow and Archimedes' Point. New York: Oxford University Press.

Price, H. (1998) Two paths to pragmatism II. In R. Casati and C. Tappolet, (Eds.), European Review of Philosophy, Vol. 3, 109–147.

Putnam, H. (1981) Reason, Truth and History. Cambridge: Cambridge University Press.

Quine, W. V. O. (1953) From a Logical Point of View. Cambridge, Massachusetts: Harvard University Press.

Ramsey, F. P. (1978a) Truth and probability. In D. H. Mellor (Ed.) Foundations: Essays in Philosophy, Logic, Mathematics and Economics. London: Routledge & Kegan Paul, 58–100.

Ramsey, F. P. (1978b) General propositions and causality. In D. H. Mellor (Ed.) Foundations: Essays in Philosophy, Logic, Mathematics and Economics. London: Routledge & Kegan Paul, 133–151.

Reichenbach, H. (1956) *The Direction of Time*. Berkeley and Los Angeles: University of California Press.

Russell, B. (1917) The relation of sense-data to physics. In his *Mysticism and Logic and Other Essays*. London: George Allen & Unwin, 145–179.

7

Analogy, Causal Patterns and Economic Choice

ROBERTO SCAZZIERI

University of Bologna
Caius College and Clare Hall, Cambridge

1 Introduction

Economic choice is often represented as rational calculus, or at least as a
process that can effectively be approximated by a model of rational deliberation.[1] However, the intellectual tradition of bounded rationality has partially transformed the above conception of economic choice, and has called
attention upon features of choice such as pattern recognition, focal points
and induction by analogy[2]. It may be argued that research on bounded ra-

---

[1] A comprehensive critical reconstruction of the aims and scope of the rationality hypothesis
in economic theory may be found in Arrow (1987). This essay is remarkable for its consideration of the theoretical embedding structure encompassing the rationality hypothesis, and
for its view that rationality 'is most plausible under very ideal conditions', so that 'when these
conditions cease to hold, the rationality assumptions become strained and possibly even self-
contradictory' (Arrow 1987, p. 69). In particular, Arrow calls attention upon the role of insti-
tutional constraints and argues that, as soon as market power and market incompleteness are
recognised, the alleged simplicity of a 'rational' decision problem is bound to break down
(see Arrow 1987, pp. 71-74).

[2] Pattern recognition is central to the conceptual framework of bounded rationality developed
by Herbert Simon and other scholars (Simon 1982 and 1983; Koffka 1935; Tversky and
Kahnemann 1974). Focal points have been defined as the focus of convergent expectations in
a classical contribution by Thomas Schelling (1960) and have more recently been considered

Stochastic Causality.
Maria Carla Galavotti, Patrick Suppes and Domenico Costantini (eds.).
Copyright © 2001, CSLI Publications.

tionality has shifted the paradigm of economic choice from the epistemic framework of indirect knowledge derived from data through utilisation of inferential methods (a model of 'vertical induction') to a different epistemic framework characterised by a combination of direct knowledge and indirect knowledge derived from the utilisation of a criterion of likeness or verisimilitude (a model of 'lateral induction'). The purpose of this essay is to investigate the implications of the above epistemic shift as to the internal structure of economic decisions. In particular, section 2 considers ways in which the boundedness of rationality suggests the consideration of causal linkages based upon the identification of partial similarity between different contexts, and the transfer of 'causal connectivity' from one context to another. Section 3 investigates the relationship between patterns and surprise, and argues that economic choice under conditions of (hard) uncertainty often presupposes the ability to detect unlikely (and unforeseen) connections and to act according to a new causal hypothesis. Section 4 considers the relationship between economic choice, rationality and the likelihood of causal hypotheses, and draws the paper to its close.

2 The Boundedness of Rationality, Partial Likeness and Causal Linkages

Bounded rationality brings about a shift from computation to pattern identification[3]. The discovery that human computational capability is limited suggests that the growth of knowledge may take place through the introduction of new focal points, which can be conducive to the re-arrangement of available evidence into new and unexpected patterns. As a result, the effectiveness of knowledge can be measured by the flexibility of 'gestalt' switches rather than by the accuracy of inferential reasoning. In general terms, bounded rationality entails the existence of constraints upon human freedom of choice, as a result of constraints upon the combination of events that the human mind is capable of assessing. Computational limitations suggest that the inferential capabilities of human beings are also limited, and are associated with the existence of constraints upon the degree of con-

---

as a possible basis of social norms and cultural paradigms (Bacharach 1993; Mehta Starmer and Sugden 1994; Sugden 1995). Analogical inference is central in John Maynard Keynes's theory of probabilistic judgement (Keynes 1921) and has led to remarkable economic applications in Keynes's theory of investment (Keynes 1936, Chapter 12 and Keynes 1937), Shackle's analysis of surprise (Shackle 1974 and 1979) and Hayek's theory of catallaxy (Hayek 1988).

[3] An effective introduction to the analysis of the linkage between the boundedness of rationality and the process by which information may be arranged according to meaningful patterns is presented in Simon (1983).

tingent planning that human beings are conceivably capable to undertake. Bounded rationality makes the human mind 'constrained' as to the amount of indirect knowledge that it may derive from limited direct knowledge by means of deductive reasoning and statistical inference. In general terms, bounded rationality suggests an epistemic model in which new information primarily derives from direct knowledge and focussing devices rather than from 'long' chains of inductive reasoning (such as the pattern of reasoning that may be associated with the maximisation of subjective expected utility). In particular, a substantial enrichment of empirical knowledge often results from 'lateral' patterns of thought, which are far from irrational but suggest the possibility of rational linkages induced by the discovery of surprising connections, rather than by the systematic elaboration of observed information.

An important route of lateral induction is suggested by analogy, and by the support that analogy may provide to the growth of empirical knowledge. A characteristic feature of analogical support is its selective nature: analogy almost never derives from a comprehensive evaluation of alternatives. Rather, the analogical support of any given hypothesis stems from partial likeness of observations and is characteristically associated with the existence of 'subjective' connections, that is, of connections that the human mind can grasp only under particular (and not easily replicable) conditions. The partial likeness of different empirical observations suggests a *multiplicity* of different routes by which a general proposition may be formulated. In this way, partial likeness supports analogical reasoning, and the latter 'opens up' the growth of empirical knowledge beyond the boundaries of Humean induction (that is, induction based upon enumeration of identical occurrences).

An early statement of the critical association between partial likeness, analogy and probabilistic thinking had been suggested by Joseph Butler:

> That which chiefly constitutes Probability is expressed in the word Likely, i.e., like some true event, like it, in itself, in its evidence, in some more or fewer of its circumstances. For when we determine a thing to be probably true, suppose that an event has or will come to pass, it is from the mind's remarking in it a likeness to some other event, which we have observed has come to pass. And this observation forms, in numberless daily instances, a presumption, opinion, or full conviction, that such event has or will come to pass
>
> Butler 1834, 'Introduction', pp. 1-2

In this connection, it may be interesting to recall John Maynard Keynes's analysis of the linkage between analogy and likeness in his *Treatise on Probability*:

> In an inductive argument we start with a number of instances similar in some respects *AB*, dissimilar in others *C*. We pick out one or more respects *A* in which the instances are similar, we argue that some of the other respects *B* in which they are also similar are likely to be associated with the characteristics *A* in other unnamed cases. The more comprehensive the essential characteristics *A*, the greater the variety amongst the non-essential characteristics, and the less comprehensive the characteristics *B* which we seek to associate with *A*, the stronger is the likelihood or probability of the generalisation we seek to establish.

Keynes 1921, pp. 219-20

Keynes's argument calls attention upon a number of points that are essential to a formal investigation of likeness and analogy. In Keynes's view, the growth of empirical knowledge derives from partial likeness according to a pattern that may be described as follows: (i) empirical generalisation presupposes a certain degree of similarity between otherwise dissimilar instances; (ii) a hierarchical structure of similar characteristics may be detected, so that it is possible to distinguish between essential and non-essential (similar) characteristics; (iii) the empirical support of any given generalisation varies directly with the number of essential common characteristics, and inversely with the number of non-essential common characteristics.

For example, given the set of essential common characteristics $\{a_1, a_2, ..., a_k\}$ and the set of non-essential common characteristics $\{b_1, b_2, ..., b_m\}$, any general proposition (hypothesis) derived from consideration of n distinct instances has greater empirical support with a higher a_k and a lower b_m, and lesser empirical support with a lower a_k and a higher b_m. In other words, empirical support is stronger if we are able to identify a set of essential common characteristics that is comprehensive enough (in which case, the attribute 'essential' could be associated with a substantial subset of common characteristics), and if the set of non-essential common characteristics includes only a limited number of features. Under such conditions, the likelihood that the *A*-characteristics will be associated with the *B*-characteristics 'in other unnamed cases' is higher than under conditions in which only a small subset of essential common characteristics can be found, and a large number of common characteristics belong to the *B* set.

It may be argued that the distinction between essential and non-essential common characteristics belongs to the 'causal core' of the empirical generalisation: it highlights the relationship between certain fundamental properties of the observed instances and certain 'accessory' characteristics of the same instances. The aim of the generalisation (hypothesis) is to explain the accessory characteristics (the B-set) in terms of essential and common characteristics (the A-set). The support of the hypothesis will be greater, the fewer are the accessory characteristics forming the *explanandum* (the B-set) and the more comprehensive is the A-set of essential characteristics forming the *explanans*.

It may be interesting to note that, if the above account is followed, causality appears to be associated with empirical uniformity: the A-set may be considered as causally related with the B-set in so far as 'unnamed cases' in which the A-set is observed are also cases in which the B-set is observed, or is likely to be observed. However, empirical uniformity per se is not sufficient to single out essential characteristics (the A-set) from among the elements of the AB set of common characteristics. As a matter of fact, the available evidence only suggests that our instances are similar in some respects (the AB characteristics) and dissimilar in other respects (the C characteristics). The utilisation of empirical uniformity in a causal argument presupposes, as a critical step, the decomposition of uniformity and its utilisation in a hierarchical logical structure. An interesting causal argument is not simply an argument about the uniformity of nature (or the certainty of correlation); it is rather an argument reflecting human choice as to the 'fundamental' or 'accessory' features of empirical evidence.

The logical structure of a causal argument primarily rests upon the relationship between the A-set and the B-set (see above). However, the empirical support of any given causal argument also reflects the range of observations for which that particular argument can be used. As a result, an empirical law by which the A-set and the B-set are joined together has greater support if the C-set covers a large variety of non-common characteristics.

To sum up, Keynes's analysis of induction by analogy calls attention upon an epistemic model in which partial likeness sustains the formation of causal structures and governs their corroboration in terms of available evidence. In particular, partial likeness determines the extent of any given generalisation, that is, the extent of the class of instances in which that generalisation may be used. Such an extent may be measured by the ratio between the elements of the C-set (non-common characteristics) and the elements of the AB-set (essential and non-essential common characteristics). Extent is greater the greater the ratio of non-common to common characteristics, and it is apparent that any given generalisation will become increasingly comprehensive as we increase the elements of the C-set and/or decrease the elements of the AB-set.

Partial likeness also determines the strength of any given hypothesis, that is, the likelihood that such an hypothesis will be supported by further empirical evidence. Strength is greater the greater the ratio of essential to non-essential common characteristics, and it is apparent that any given hypothesis will be stronger as we increase the elements of the A-set and/or decrease the elements of the B-set. In other words, a strong hypothesis is one supported by a high number of common essential characteristics (leading to a powerful *explanans*), and/or one limited to a small number of common non-essential characteristics (leading to a narrow *explanandum*).

It may be desirable to consider an index measuring the over-all explanatory power of any given argument, and thus combining a measure of extent and a measure of strength. A simple intuitive index may be one constructed by considering the proportion between strength ratio and extent ratio:

$$(A/B):(C/AB), \text{ or } A^2/C.$$

The above index makes the explanatory power of any given hypothesis H to vary directly with the number of elements in the A-set (strong dependence) and inversely with the number of elements in the C-set (weak dependence). The combined explanatory power of hypothesis H is positively influenced by the number of essential common characteristics in set A and negatively influenced by the number of non-common characteristics included in set C. This outcome is not surprising, since our purpose is to measure strength relatively to extent. Our index makes explanatory power very responsive to the number of A characteristics and more weakly responsive to the number of C characteristics. Any given increase in the number of A characteristics will affect both strength and extent: it will increase strength and reduce extent (see above). However, the effect on extent will be more than matched by the effect on strength, so that the over-all explanatory power of the hypothesis will be increased. On the other hand, any increase in the number of C characteristics will increase the extent of the hypothesis leaving its strength unchanged. This leads to a diminished explanatory power of H, since strength relative to extent will be reduced.

It may be argued that an index measuring strength relative to extent is not always the most appropriate way to describe the explanatory power of a hypothesis. Why shall we think that explanatory power has diminished simply because relative strength has diminished? After all, strength has been defined independently of extent, and greater extent reflects the increased variety of our sample space.

A possible way to deal with the above criticism could be to consider a different index of explanatory power, which may be constructed by multiplying a measure of strength by a measure of extent. This combined index could be interpreted as a measure of strength weighted by extent (or as a measure of extent weighted by strength):

$$(A/B) \cdot (C/AB), \text{ or } C/B^2.$$

The above index makes the explanatory power of any given hypothesis H to vary directly with the number of C characteristics, and indirectly with the number of B characteristics. An intuitive interpretation could be that a greater number of C characteristics exert a positive influence on explanatory power, by increasing the extent of the hypothesis. On the other hand, a greater number of B characteristics exert a negative influence on explanatory power by reducing the strength of the hypothesis. It is worth noting that, due to the particular weights adopted, the influences of A characteristics on strength and extent respectively cancel each other out, so that explanatory power appears to be independent of the number of common and essential characteristics (A characteristics). The critical elements influencing explanatory power will be the number of non-common characteristics (a measure of variety) and the number of common non-essential characteristics (the B characteristics forming the *explanandum*). In this way, the explanatory power of any given hypothesis will be assessed by considering a combined index measuring the scope of the explanation being sought and the variety encompassed by the sample space.

This second index is liable to criticism on the ground that it comes to reflect the range of relevant phenomena (as measured by C) and the scope of the explanation being sought (as measured by B), to the exclusion of other features, such as the ratio of essential to non-essential common characteristics. As a result, it could be argued that the second index only associates explanatory power with the variety of the sample space and the 'dimension' of the *explanandum*, while a more direct measure of likelihood is excluded.

A possible way to deal with this problem could be to consider a third index, which may be constructed by weighing index C/B^2 with our measure of explanatory strength A/B:

$$(C/B^2) \cdot (A/B) = (AC)/B^3.$$

The above index would measure the explanatory power of any given hypothesis H by assigning a certain degree of influence to the A, B and C characteristics. However, A and C characteristics only exert a relatively weak influence upon explanatory power, which is strongly responsive to the number of B characteristics (common and non-essential characteristics).

Perhaps the structure of our third index reveals a property of causal linkages established by means of analogical induction, and suggests that a 'manipulation' of causality targeting the strength of a causal connection or the variety of the sample space only exerts a weak influence upon explanatory power. On the other hand, explanatory power is much more responsive to a manipulation targeting the number of B characteristics, and thus influencing what may be called the 'focus' of a causal connection. In other words, a manipulation reducing the number of common and non-essential characteristics entering the *explanandum* is much more effective in increasing the explanatory power of any given hypothesis relatively to a manipulation increasing the range of common and essential characteristics, or increasing sample variety. A manipulation targeting the focus of the *explanandum* appears to be the most effective way to increase the likelihood of any given hypothesis.

The above argument also suggests that causal linkages associated with analogical induction may be based upon a 'lateral' selection of theories, whereby the explanation of an unexpected occurrence (which is to be proposed by transferring a certain theory from one context to another) is made possible by modifying the structure of the *explanandum* and thus by targeting its focus in a different way. In a sense, causality associated with analogical induction makes use of theories about the world, but presupposes the ability to select appropriate theories as the set of common and non-essential characteristics (the B set) gets modified. For a change in the B set entails a different pattern of partial similarities across the constituents of empirical evidence, and suggests the utilization of different theories. In this case, rational inferences come to resemble inferences grounded upon good reasons and cognitive skills, rather than arguments based upon an explicit set of formal rules. As a result, identification of causality would presuppose theoretical pluralism and sensitivity to often minor changes in empirical evidence [4].

---

[4] As in data warehouses, this inferential structure is based upon a collection of theories and presupposes the ability to switch from one theory to another as the empirical structure is changed. The recent work on so-called 'materialized views', their role in data mining and the conditions for their self-maintenance has addressed precisely the issue of the sensitivity of multiple theories confronted by a change in empirical structure. (See, for example, Liang, Li, Wang and Orlowska 1999; Mohania and Kambayashi 2000.)

3 Patterns and Surprise

Surprise is associated with the unexpected, and attempts have been made to obtain a measure of surprise in terms of probability: if any given event is known before its occurrence, then surprise will be greater as its probability is lower (see also Edwards 1992, p. 202; 1st edn. 1972). However, it may be argued that 'the quantification of surprise in terms of probability is likely to tell only half the story' (Edwards *ibidem*, p. 203). And the reason for this could be that 'it seems as though we each carry around a mental picture of what the world will be like in a few moments' time, based on our knowledge of the past and the present. We are surprised about events which are not part of this picture, even though, had we taken more thought, they might have been' (Edwards *ibidem*, p. 203).

In this connection, the relationship between patterns and surprise is worth exploring. As a matter of fact, the above argument entails that patterns, considered to be connecting principles by which experience can be organized and interpreted, can exert a critical influence upon the apprehension of likeness and the formulation of explanation (identification of causality). In particular, any given pattern (mental picture) could be associated with a particular explanation of 'the unexpected' and a certain measure of surprise (see above). However, occurrence of an event that was not at all anticipated (in terms of existing patterns) could either appear mysterious or suggest a different pattern. In the latter case, a new explanation may be attempted that would critically depend upon the ability to conceive of a space of events in which the new occurrence is possible. This case, in terms of the conceptual framework discussed in section 2 above, could be one in which the number of dissimilar characteristics C is increased, and yet the B-set can be explained in terms of the A-set. For example, our surprise could derive from the fact that the A-set has not generated the B-set, or from the fact that certain B characteristics have been generated even in the absence of the (complete) A-set. A new pattern encompassing the unanticipated occurrence could result, respectively, from increasing the number of A characteristics while simultaneously reducing the number of B characteristics, or by increasing the number of B characteristics while simultaneously reducing the number of A characteristics. In the former case, the variety of the A-set is increased while the variety of the B-set is diminished: there are more essential characteristics, and less common non-essential characteristics. This may account for the unanticipated failure of the A-set to generate the B-set. In the latter case, the variety of the B-set is increased, while the variety of the A-set is reduced: there are less essential characteristics and more common non-essential characteristics. It follows that a *B*-set can be generated even if not all the *A* characteristics have been observed.

It may be argued that the analytical treatment of surprise is especially conducive to induction by identification of partial likeness (analogical induction), and to explanation by virtue of the identification of imagined possible outcomes. In this connection, surprise is closely associated with choice, particularly if the latter is conceived as the result of a process in which the chooser's mental frame exerts a critical influence. As G.L.S. Shackle has noted, choosing presupposes the consideration of rival options. And such options 'are the products of [the chooser's] original thought, his re-assembly, in configurations never realized in the field in its past, of the elements into which he resolves his basic scheme of that field. These re-assemblies can be rivals for the same stretch of time-to-come, but they must conform to the scheme of Nature and of human nature which he has evolved, [...] they must be possible in his judgement' (Shackle 1979, p. 13).

Hard uncertainty may be conceived as a situation in which, given a particular event, we are unable 'to choose between the many probability models in which we could embed the particular event' (Edwards 1992, p. 20). It may be argued that hard uncertainty (as defined above) makes it difficult to conceive rational human choice (in general) and economic choice (in particular) as the outcome of deliberation in terms of a particular probability model. It follows that a state of hard uncertainty suggests a different route to rationality, and makes economic choice to reflect a criterion similar to the 'surprise setting' described above. In particular, choosers confronting an unanticipated event will not be able to select a probability model embedding that event, and will fall back to pattern identification (a discovery procedure) in the attempt to give reasons for the unexpected occurrence. Hard uncertainty will be conducive to search for patterns by which likeness may be detected and rational decisions be taken. As a result, the rationality of economic choice can be preserved, but the grounds of rationality are shifted from calculus to design. Rationality of choice no longer depends upon a calculus of numerical probability (that is itself based upon a known probability distribution), but reflects the chooser's skill in distinguishing between essential and non-essential characteristics of occurrences and in identifying a causal hypothesis, which may in turn suggest a definite probability model in which events under consideration could be embedded[5].

---

[5] It is worth noting that the above conception of rational choice is akin to Adam Smith's view of human reason in terms of imagination and prudence (Smith 1759; see also Porta and Scazzieri 2001), and to John Maynard Keynes's account of rational judgement in terms of a balance between the probability of any given proposition and the weight of the corresponding

Rationality of economic choice, under conditions of hard uncertainty, is primarily a science of surprise. Choosers are confronting unanticipated events that are not easily embedded by existing probability models, so that rationality presupposes the ability to make distinctions (between essential and non-essential characteristics) and to 'reduce' the unanticipated outcome to a new causal hypothesis[6].

In this perspective, economic rationality presupposes serendipity, that is, a set of connecting principles by which choosers are able to make 'happy' discoveries 'by accidents and sagacity, of things they were not in quest of' (Walpole 1834, ms 1754). In this connection, causality appears to be a critical step in rational deliberation under hard uncertainty, when the chooser may be unable to select between rival probability models. But serendipity (the ability to make happy surprises possible) presupposes imagination, which is the chooser's ability to make a new distinction between essential and non-essential characteristics, and to arrange them in a large field of possible combinations. As a result, a connection emerges between rationality and freedom: rationality (as a propensity to introduce new distinctions and associations) enables human beings to make 'happy discoveries' (see above) under conditions of hard uncertainty; freedom (as a condi-

---

argument (see Keynes 1921; see also Keynes 1936, Chapter xii, and Keynes 1937). In Keynes's view, the rationality of choice under conditions of hard uncertainty is dependent both on the ability to assess probability distributions in terms of the available evidence, and on the ability to identify which evidence is relevant (and reliable) for any given problem of choice: 'it would be foolish, in forming our expectations, to attach great weight to matters which are very uncertain. It is reasonable, therefore, to be guided to a considerable degree by the facts about which we feel somehow confident, even though they may be less decisively relevant to the issue than other facts about which our knowledge is vague and scanty' (Keynes 1937). As noted by G. L. S. Shackle (1974), Keynes's account of rational economic choice under conditions of hard uncertainty shows special attention for the mutable character of relevant expectations, and for the focussing devices used by economic agents in order to make a decision possible: 'the expectations, which together with the drive of needs or ambitions make up the "springs of action", are at all times so insubstantially founded upon data and so mutably suggested by the stream of "news", that is, of counter-expected or totally unthought-of events, that they can undergo complete transformation in an hour or even a moment, as the patterns in the kaleidoscope dissolve at a touch. [Human beings] are conscious of their essential and irremediable state of un-knowledge and [they] usually suppress this awareness in the interest of avoiding a paralysis of action' (Shackle 1974 p. 42).

[6] It may be interesting to note, in this connection, that the introduction of a new causal hypothesis, by which the unanticipated outcome (surprise) may be accounted for (and a new probability distribution may be generated) could be an instance of 'incidental causation', that is, an unintended result of a distinction (between essential and non-essential characteristics) that has no such purpose in view.

tion influencing the range of human choice) is enhanced by the ability to generate new mental frames, and thrives when serendipity is possible[7].

4   A Theory of Choice, Distance and Likelihood

Economic choice is generally associated with rational calculus and the identification of a 'best' course of action. However, issues relative to the framework of choice and the history of choosing are often left aside. In the first case (framework of choice), the economic analysis of choice is carried out by overlooking pattern identification, so that the process by which the chooser identifies the space of possible (and relevant) configurations is not explicitly considered. In the latter case (history of choosing), economic analysis seldom investigates the thought processes by which choosing is related with a changing mental representation of the possible (and relevant) states of the world.

The foregoing argument suggests that economic choice is to a large extent a cognitive process shaped by mental frames and pattern identification. It also suggests that the history of choosing cannot be reduced to an intertemporal sequence of 'best' decisions, or to a calculus of 'best' decisions directly stemming from a changing informational endowment. The process of economic choice is primarily a process by which effective frameworks are generated and modified over time as a result of unanticipated outcomes (surprises). As argued above, economic choice is to a large extent a process associated with the rational handling of unanticipated outcomes under conditions of hard uncertainty (as defined above). If the chooser is unable to select a relevant probability model, she/he will have to reconsider his/her configuration of possible and relevant options. It may be argued that a rational criterion by which a new mental frame is generated would be one in which the unanticipated occurrence comes to be explained, and a probability model is selected in which that particular event can be embedded.

In particular, rational choice comes to resemble the classical pattern of good reasons combined with satisficing (or standardized) rules of behaviour. Focal points, which may be defined as the foci of convergent expectations at the level of social interaction (see Schelling 1960, Chapter iv), suggest a standard (or a set of standards) for the choosing process, and identify points of reference *both* when a strategy of action and a particular theory about the world have to be selected. The formal symmetry between strategy

---

[7] The above account of the relationship between serendipity and freedom may be related with the possibility to outline an analytical representation of freedom in terms of the structural diversity of opportunities within a stochastic process (see Suppes 1996 and 1997).

choice and theory choice (see also Rubinstein and Zhou 1999, pp. 205-6) highlights the cognitive dimension of economic choice (more generally, of choice relative to a course of action); at the same time, it also brings to the fore the 'singular' (or 'historical') dimension of theoretical choice , that is, the fact that choice of a particular theory about the world comes under the influence of circumstances that generally make it path-dependent and not replicable.

It may be argued that a focal point influences strategy or belief choice by means of what has been called 'minimal distance' criterion (see Yu 1973; Rubinstein and Zhou 1999). In this case, the chooser's solution would be the strategy (or, respectively, the belief) by which the distance from the status quo option (or, respectively, from the currently accepted set of beliefs) is minimized.

The rationality of choice guided by focal points is bounded and path dependent[8]. In this connection, it is worth examining some implications stemming from the theory of partial similarity and induction outlined in the previous sections. As noted above (see section 2), the explanatory power of any given hypothesis comes to reflect the focus of the corresponding causal connection, and the latter may be increased by reducing the number of common and non-essential characteristics (B characteristics) entering the *explanandum*.

At the same time, a reduction in the number and type of B characteristics may influence the pattern of partial similarities shown by empirical evidence and enhances the focus of a causal hypothesis. For example, a shift from set $\{b_1, b_2, ..., b_m\}$ to set $\{ b^*_1, b^*_2, ..., b^* \}$, with $k < m$ and $b_i = b^*_i$ except for $b_1 \neq b^*_1$ and $b_m \neq b^*_k$, induces at the same time a change in partial similarities across empirical evidence and a better focus of the associated causal hypothesis (or hypotheses). The likelihood of the new causal hypothesis (hypotheses) may be higher than the likelihood of the original hypothesis (hypotheses) due to contraction in the number of B characteristics, which sharpens the focus of partial similarities without increasing the irrelevant variety of the sample space (this is associated with the number of C characteristics, that is, with the number of characteristics that are different from one occurrence to another).

The above framework suggests that lateral thought (induction by analogy) may be essential in the generation of a new causal hypothesis, and that the likelihood of a successful hypothesis may be enhanced by suitable ma-

---

[8] An interesting proposition proven by Rubinstein and Zhou is that 'the minimal distance function does not satisfy a path independence axiom' (Rubinstein and Zhou 1999, p. 206). In other words, the order in which new information is added may influence the process by which a strategy is revised or, respectively, a belief-set is updated.

nipulation of partial similarities. The central point here is that under certain conditions a change in the structure of evidence is associated with a change in the number and type of B characteristics, such that partial similarities get reduced and better focussed. This suggests that successful learning (theory updating and theory change) may be critically dependent upon the ability to explore a space of partial similarities and to concentrate one's attention upon a limited number of similar characteristics.

The formal symmetry between theory choice and strategy choice (see above) suggests a number of interesting implications as far as economic choice is concerned. In particular, the rationality of choice can be measured, at any given time, by the likelihood $L(H|R)$ of conditional proposition H, which may alternatively be considered as a theory about the state of the world or as a successful project about human conduct. The likelihood of a theory describes how good that theory is as an approximation to truth; the likelihood of a successful project describes how effective that project could be as an approximation to the satisfaction of human preferences.

Mental frames influence both theory choice and strategy choice (see above). In particular, the structure of the empirical occurrence to be explained (or the structure of the action to be determined) reflects the nature and type of relevant characteristics. For example, the distribution of characteristics between classes A, B and C influences the possible pattern(s) of partial similarities and thus also the focus of theory choice (or strategy choice). A change in the structure of empirical evidence, or in the structure of required action(s), may change the focus of theory (or the focus of strategy), and thus alter in a significant way the degree to which a particular theory, or a particular strategy, can be a successful one[9].

Economic choice as rational choice under conditions of hard uncertainty may be described as a process by which effective hypotheses (or strategies) can be generated and unsuccessful hypotheses (or strategies) can be discarded. Hard uncertainty makes it impossible to select a particular theory (or strategy) by considering a known probability distribution. However, the process by which a change in the structure of empirical evidence leads to a change of hypothesis (or to a change of strategy) may be associated with a transformation in the relevant space of events, which may result either from the introduction of more detailed language (language increment) or from the introduction of more extensive information (information

---

[9] As argued above, a contraction in the number of B characteristics may be expected to sharpen the focus of choice, thus increasing the likelihood of theory or the effectiveness of strategy.

increment)[10]. In both cases, there is a change in empirical structure, which may induce a different assessment of probabilities and enhance (or reduce) the likelihood (effectiveness) of received hypotheses (strategies).

To a large extent, rationality of choice can be measured, at any given time, by the likelihood $L(H|R)$ of the conditional proposition H that the chooser adopts given the set R of occurrences, and the rationality of the process by which new hypotheses (or strategies) are generated may be associated with the effectiveness of each new hypothesis (or strategy) as a response to surprise (and thus to a new configuration of the relevant space of events). It follows that rationality of choice under conditions of hard uncertainty (when no probability distribution is known) is radically different from rationality of choice on the basis of a known (or assumed) probability distribution. In the former case, 'a specific model is coupled with considerations of long-run frequency so that events of equal probability must be *predicted* equally' (Edwards 1992, p. 217); in the latter case, the agent's focus is shifted from probability distribution to space of events, and the cognitive issue of analytical representation must be approached. It has been argued that hard uncertainty (surprise) entails 'a tension between the probability model and the sample space' (Edwards 1992, p. 217). Rational choice, in this case, is primarily the ability to switch to a new mental frame, that is, the ability to generate a new space of events that assigns high probability to the surprising sample. However, the *transition* from one mental frame to another is associated with likelihood rather than probability, for the relevant standard in assessing alternative frameworks (alternative empirical structures) 'is a relative probability only, suitable to compare point with point, but incapable of being interpreted as a probability distribution over a region, or of giving any estimate of absolute probability' (Fisher 1912, as quoted in Edwards 1992, p. 227).

This essay has argued that economic rationality under conditions of hard uncertainty is intrinsically associated with the chooser's ability to give reasons for unanticipated occurrences (surprises), her/his ability to discard unsuccessful hypotheses (strategies), and the effectiveness by which unanticipated occurrences give rise to new and interesting patterns of partial similarities. It has also been argued that the ability to identify a new pattern (a new empirical structure), such that the surprising sample takes high

---

[10] The structure of events presupposed by a theory of probability is examined in Crisma (1988). In particular, Crisma highlights the objective features of any given space of events, when the latter is construed from shared language and informational endowment. This approach entails that a change in the structure of events (or in the representation of events) may arise from an 'increment of language' (Crisma 1988, pp. 21-22), an 'increment of information' (Crisma 1988, pp. 22-24), or both. I am grateful to Attilio Wedlin for calling my attention to Crisma's contribution.

probability, may reflect the agent's ability to see lateral associations, to discard irrelevant variety, and to focus upon a powerful causal hypothesis (or an effective strategy).

Acknowledgements

I wish to express my gratitude to the participants of the Bertinoro Conference for enlightening discussions and the general intellectual climate. In particular, I am grateful to Patrick Suppes, Alessandro Vercelli and Attilio Wedlin for the long conversations in which I had an opportunity to test the original formulation of my ideas and to see some of their implications. I am also grateful to Anthony Edwards for his encouragement to consider the relationship between likelihood and surprise. It goes without saying that I am solely responsible for the views expressed in this essay.

References

Arrow, K. J. (1987) Economic theory and the hypothesis of rationality. In J. Eatwell, M. Milgate and P. Newman (Eds), *The New Palgrave. A Dictionary of Economics*, Vol 2. London: Macmillan, 69-75.

Bacharach, M. (1993) Variable universe games. In K. Binmore, A. Kirman and P. Tani (Eds.), *Frontiers of Game Theory*. Cambridge, Mass.: MIT Press, 255-75.

Butler, J. (1834) *The Analogy of Religion, Natural and Revealed, to the Constitution and Course of Nature,* with a Preface by Samuel Halifax. London: Longman and Co. (First edn 1736.)

Crisma, L. (1988) Dalla certezza all'incertezza: aspetti dinamici in una impostazione soggettiva. In *Atti del Convegno su incertezza ed economia* (Trieste, 29-30 ottobre 1987), Trieste: Edizioni LINT, 11-46.

Edwards, A. W. F. (1992) *Likelihood*. Expanded Edition, Baltimore and London: The Johns Hopkins University Press (1st edn 1972).

Fisher, R. A. (1912) On an absolute criterion for fitting frequency curves. *Messenger of Mathematics*, 41, 155-60.

Hayek, F. A. (1988) *The Collected Works of Friedrich August Hayek*, vol. I, edited by W. W. Bartley. London: Routledge.

Keynes, J. M. (1921) *A Treatise on Probability*. London: Macmillan.

Keynes, J. M. (1936) *The General Theory of Employment, Interest and Money*. London: Macmillan.

Keynes, J. M. (1937) The general theory of employment, *Quarterly Journal of Economics*, 51, 209-23.

Koffka, K. (1935) *Principles of Gestalt Psychology*, London: Routledge and Kegan Paul.

Liang, W., Li, H., Wang, H., and Orlowska, M. E. (1999) Making multiple views self-maintainable in data warehouse. *Data and Knowledge Engineering*, 30, n. 2, 121-134.

Mehta, J., Starmer, C., Sugden, R. (1994) The nature of salience: An experimental investigation of pure coordination games. *The American Economic Review*, 84, n. 3, 658-73.

Mohania, M. and Kambayashi, Y. (2000) Making aggregate views self-maintainable. *Data and Knowledge Engineering*, 32, n.1, 87-109.

Porta, P. L. and Scazzieri, R. (2001) Coordination, connecting principles and social knowledge: An introductory essay. In P. L. Porta, R. Scazzieri and A. Skinner (Eds), *Knowledge, Social Institutions and the Division of Labour*. Cheltenham, UK and Northampton, MA, 1-32.

Rubinstein, A. and L. Zhou (1999) Choice problems with a 'reference' point. *Mathematical Social Sciences*, 37, 205-9.

Schelling, T. C. (1960) *The Strategy of Conflict,* Cambridge, Massachusetts: Harvard University Press.

Shackle, G. L. S. (1974) *Keynesian Kaleidics. The Evolution of a General Political Economy*. Edinburgh, Edinburgh University Press.

Shackle, G. L. S. (1979) *Imagination and the Nature of Choice*. Edinburgh: Edinburgh University Press.

Simon, H. (1982) *Models of Bounded Rationality*, volumes i and ii, Cambridge, Mass.: MIT Press.

Simon, H. (1983) *Reason in Human Affairs*. Stanford: Stanford University Press.

Smith, A. (1759) *The Theory of Moral Sentiments*. London: A. Millar; Edinburgh: A. Kincaid and J. Bell.

Sugden, R. (1995) A theory of focal points. *The Economic Journal*, 105, 533-50.

Suppes, P. (1996) The nature and measurement of freedom. *Social Choice and Welfare*, 13, 183-200.

Suppes, P. (1997) Freedom and uncertainty. In H. G. Natke and Y. Ben-Haim (Eds), *Uncertainty: Models and Measures, Mathematical Research*. Berlin: Academie Verlag, 69-83.

Tversky, A. and Kahnemann, D. (1974) Judgement under uncertainty: Heuristics and biases. *Science,* 185, 1124-31.

Walpole, H. (1834) Letter to Mann, 28 January 1754, in H. Walpole, *Letters to Sir H.Mann* […] *British Envoy at the Court of Tuscany* […], 3 volumes, London, R. Bentley, vol. iii.

Yu, P. L. (1973) A class of solutions for group decision problems. *Management Science*, 19, 936-46.

8

Epistemic Causality and Hard Uncertainty: A Keynesian Approach

ALESSANDRO VERCELLI

University of Siena

1 Introduction

The interplay of epistemic and empiric conditions of human behaviour plays a crucial role in economic causality but it is not satisfactorily analysed by the existing approaches to economic causality, including the most influential of them: Granger causality. In order to find a more satisfactory approach to economic causality, this paper draws inspiration from the contributions of Keynes. In particular it is argued that the systematic use of probabilistic causality in the *General Theory* is deeply rooted into the theory of probabilistic causality outlined in the *Treatise on Probability* which is based on the crucial distinction between epistemic and empiric causality. In this view, since epistemic causality is conceived as probabilistic, also empiric causality has to be conceived as probabilistic in economics since in this case, differently than in natural sciences, the observed events are mediated in a crucial way by epistemic links, such as beliefs and expectations. In particular, the awareness of the relevant ignorance, represented by what is called by Keynes the 'weight of argument', affects in a crucial way the behaviour of economic agents. The Keynesian approach is extended by observing that a change in the weight of causal arguments represents a sort of

Stochastic Causality.
Maria Carla Galavotti, Patrick Suppes and Domenico Costantini (eds.).
Copyright © 2001, CSLI Publications.

'second-order epistemic causality' that may become a possible independent source of empiric causality.

The approach to causality that has prevailed in economics in the last two decades is based on a peculiar definition of probabilistic causality suggested by Granger (1969) who elaborated on a previous hint by Wiener (1958). The basic idea is very simple: a stochastic variable x_t causes in the Granger sense the stochastic variable y_t whenever the knowledge of the past and present values of x_t helps forecasting the future values of y_t as it reduces the variance of the prediction errors (in some well-specified sense). Though the technical details of the definition and of its implementation through econometric tests are quite sophisticated, the underlying conceptual framework is very simple, if not simplistic, as it is nothing but a particularly naive version of Humean causality: the stochastic variable x_t 'causes' (in the Granger sense) the stochastic variable y_t whenever we may detect in their co-movements a significant correlation and a prevailing time lag between their 'realisations' that specifies the direction of the causal arrow.

This new concept of causality has not been immediately successful because most economists were at the time suspicious of any concept of asymmetric causality apparently inconsistent with the general interdependence of economic variables as represented in general equilibrium theory, while the few economists dealing with asymmetric causality were utilising concepts of deterministic causality,[1] and were not prepared to recognise the epistemological legitimacy of any concept of probabilistic causality. However since the late 1970s the Granger concept of causality began to spread and to be routinely applied in economic analysis ousting almost completely the other concepts of causality and obtaining a success that no concept of causality had obtained before in economics. Among the main reasons of the unprecedented success of Granger causality the following reasons may be briefly recalled:

- The strong affinity with the new emergent stream of economic analysis based upon the assumption of rational expectations, both in the modelling techniques utilised (rooted in the theory of stochastic processes) and in the underlying vision of the real world conceived as closed and stationary.[2]

---

[1] The concept of causality suggested by Simon (1952), the most popular within economics until the late 1970s, was based on the order of computation of the endogenous variables implicit in the structure of coefficients which characterises a system of equations.

[2] We define the world as *closed* whenever the decision maker knows all the possible states, all the possible acts and the probability of occurrence of all the possible consequences of these acts in any possible states. We define the world as *stationary* whenever it is assumed that the set of possible states, acts and consequences do not change through time, and the same act leads to the same consequence conditional to the same state with the same probabil-

- The wealth of apparently simple econometric tests to corroborate or falsify causal assertions based upon the Granger concept so that the new kind of economic analysis could distinguish a priori between endogenous and exogenous variables, a necessary requisite for its correct implementation.

- The philosophical legitimacy of the Granger concept of probabilistic causality based on its formal analogy with philosophical concepts such as that of Suppes (1970) which in the meantime had acquired prominence in the philosophical debate.

Notwithstanding its extraordinary success, the approach of Granger causality continued to raise many sharp criticism (for a critical survey see, e.g., Vercelli 1989, 1991, and 1992). We just recall here two of them which are particularly relevant for this paper:

- Granger causality is correctly defined as relative to a set of background information; however it is claimed by its practitioners that such a set only includes the past and present values of the relevant stochastic variables while, differently from the other causal concepts, it does not need to include any kind of a priori theoretical knowledge. Therefore it is claimed that the causal assertions grounded on the results of Granger causality tests are unconditional to a priori theoretical assumptions and that this makes this approach superior to the competing alternative causal approaches. Unfortunately it is possible to demonstrate that this claim is false (see Vercelli, ibidem) and that this undermines any general claim of superiority for the Granger approach vis-à-vis alternative approaches.

- The Granger concept of probabilistic causality implies by definition the existence of some sort of relevant uncertainty in the economic system; however both the conceptual underpinnings of this approach (in particular the crucial link with prediction rather than with explanation) and the assumptions of the tests (that in particular have to rely on the stationarity of the relevant stochastic processes) imply that it can be applied only to situations characterised by a very weak kind of uncertainty which we are going to call 'soft' uncertainty.[3]

---

ity. Though these two properties attributed to the real world are very demanding they are explicitly assumed in the received decision theory either objectivist (Morgenstern-von Neumann 1944) or subjectivist (Savage 1956), which underlies both orthodox economic theory and Granger causality.

[3] In this paper we will distinguish between *soft* uncertainty, whenever the beliefs of the decision-makers may be represented through a unique, fully reliable, additive probability distribution and *hard* uncertainty whenever the beliefs of the decision-makers may be represented

These two shortcomings are strictly linked, as they are both a symptom of a narrow and simplistic way of conceiving the relationship between the epistemic and empiric conditions of causality. The assumption of rational expectations which characterises the style of economic analysis within which Granger causality has become so popular is an extreme, though ubiquitous, example of this attitude: the crucial assumption that characterises the 'rational expectations hypothesis', i.e., that the subjective probability distribution coincides with the objective probability distribution, by definition eliminates any tension, indeed operational distinction, between epistemic and empiric conditions of causality and decision.

In order to work out a more satisfactory notion of probabilistic causality based on a more sophisticated analysis of the complex relationship between epistemic and empiric conditions of causality we may still find useful inspiration in the contributions by Keynes. Though the passages explicitly dedicated to causality by Keynes in both the *Treatise on Probability* (1921, from now on *TP*) and its economic works are sparse and scanty, still they are inserted in a very rich conceptual framework which makes them a pregnant source of inspiration. The *TP* hints at a notion of probabilistic causality which, at first sight, may be considered as an early forerunner of the Granger notion (Vercelli 1989). However, though the formal language utilised by Keynes is obsolete and much less sophisticated than that of Granger, on the contrary its philosophical underpinnings are much sounder from the epistemological point of view. In particular Keynes emphasises that 'since our knowledge is partial, in the use of the term *cause* there is always an explicit or implicit reference to a limited corpus of knowledge' (*TP*, p. 306) This conceptual framework has to be made explicit ex ante as clearly as possible in order to clarify the meaning and scope of any causal assertion. In particular, in the absence of such a conceptual framework, we could not distinguish between a spurious and a genuine cause (see Vercelli 1989). The claim of superiority of Granger causality based on its alleged unconditionality to a priori theoretical assumptions must therefore be rejected, if not reversed. The same opinion may be found already in the *TP*: 'The opposite view, which the unreliability of some statisticians has brought into existence—that it is a positive advantage to approach statistical evidence *without* preconceptions based on general grounds, because the temptation to 'cook' the evidence will prove otherwise to be irresistible— has no logical basis and need only be considered when the impartiality of an investigation is in doubt' (*TP*, p. 338). This point is a fundamental

---

only through a non-additive probability distribution or through a plurality of probability distributions that may be additive but none of which is considered as fully reliable (on this distinction and its implication for decision theory see Vercelli (1999)).

investigation is in doubt' (*TP*, p. 338). This point is a fundamental assumption which underlies the following analysis. In addition Keynes has a very sophisticated understanding of the nature and implications of uncertainty for the 'moral sciences' (i.e., human and social sciences) such as economics, as well as of the complex interaction between epistemic and empiric conditions in decision making and causality (Keynes 1939, and *Collected Writings* (from now on *CW*) XIV and XXIX). In particular he emphasises the need of distinguishing different modalities of uncertainty which have different implications for rational decision making; one important implication underlined by Keynes is the fact that probabilistic arguments may have a different 'weight' that is liable to affect the rational behaviour of decision makers.

This paper intends to clarify the nexus between epistemic and empiric conditions of probabilistic causality within a Keynesian conceptual framework applied to the economic empirical field. In the second section the contributions by Keynes to probabilistic causality are briefly summarised and discussed by explicitly connecting the contributions put forward in the *TP* with those contained in his economic works (with special reference to the *General Theory*, 1936, from now on *GT*). In the third section the relationship between the weight of argument and probabilistic causality is discussed; this suggests the introduction of a new concept here christened as 'second-order epistemic causality'. In the 4th section an example drawn from the *GT* illustrates the pragmatic relevance not only of the Keynesian concepts of epistemic causality and weight of argument but also of the concept here introduced of 'second-order epistemic causality'. A few concluding remarks follow.

2   Epistemic Causes and Economic Causes in Keynes

In the TP there is an unsolved tension between *causa cognoscendi*, which may be translated as 'epistemic cause', and *causa essendi* which could be translated as 'ontological[4] cause' according to the philosophical tradition to which the Latin words alluded to, but that we prefer to translate as 'empiric cause' in order to avoid in this paper philosophical problems which are not central for the argument. In particular, the empiric cause is seen as deterministic, i.e., its occurrence is seen basically as a necessary and/or sufficient condition ceteris paribus for the occurrence of the effect, while the epis-

---

[4] Of course, also, the epistemic conditions have their 'being', which is not necessarily altogether different from that of natural phenomena, unless an extreme dualism such as that of Descartes is assumed. In our dichotomy 'empiric' stands for 'directly observable' while the epistemic states are not directly observable.

temic cause is seen as probabilistic since we are led, according to Keynes, to use the term 'cause' in a broader sense than that of 'sufficient cause' or 'necessary cause', because being rarely evident the necessary and/or sufficient causal link of particular events with particular events, the strict sense of the word is almost useless (*TP*, pp. 306-7), particularly in moral sciences (Keynes 1939). The ontological determinism of the empiric cause is for Keynes just the 'received view' on the natural world apparently confirmed by the amazing empirical success of Newtonian physics but he is fully aware that the limits of the human mind coupled with the complexity of the real world allow in most cases only a knowledge of probabilistic relations, as Laplace had clarified already long ago. This point of view leads Keynes to conceive of the epistemic cause as probabilistic even when the underlying ontological cause is assumed to be deterministic. This is true also for the physical world with limited and partial exceptions in the fields, such as celestial mechanics, where the phenomena to be forecasted are particularly simple (as a solar or lunar eclipse) and the relevant knowledge that may be acquired is almost complete for the purpose, so that a deterministic version of epistemic causality may be applied. However, the predominance of the probabilistic version of epistemic causality is much more evident in the 'moral sciences' where the phenomena are particularly complex and crucially depend on the subjectivity of decision-makers. This point which is just hinted at in the *TP* is developed on many occasions in his economic works. However the only definition given by Keynes of the probabilistic version of epistemic causality may be found in the *TP*. Translating his definition in the modern language suggested by Suppes (1970) we may reformulate Keynes's definition in the following way:

A_t is an epistemic cause of B_t relative to Z_t iff:
(i) $P(A_t \cap Z_t) > 0$
(ii) $P(B_{t'} \mid A_t \cap Z_t) \neq P(B_{\varepsilon'} \mid Z_t)$, $t' \geq t$,

where A_t and $B_{t'}$ are events occurring respectively at time t and t', Z_t is the background information (including the relevant theoretical assumptions) and P the probability of occurrence of the events under the specified conditions.

This definition is substantially equivalent to the definition of 'potential prima facie' cause in Suppes (1970) apart from two minor differences:

- A temporal lag between the occurrence of the cause and the effect is not required, and thus the possibility of contemporaneous causation is explicitly admitted;

- The occurrence of the cause may reduce the probability of occurrence of the effect; in other words the possibility of an inhibitory cause is explicitly admitted.

In the *GT* Keynes clarifies that probabilistic *epistemic* causes may play the role of *empiric* causes in economics (as well as in other human and social fields). A change in the quality and quantity of the relevant knowledge on the part of the decision-maker is likely to modify, even significantly and abruptly, the expectations over the variables which affect decisions. Therefore, generally speaking, in the economic field, as well in the other human and social fields, even the empiric causes may be, and typically are, probabilistic. We have a chain of causes which goes from the new evidence available to the change of expectations (epistemic cause) that determine a change in decisions which produces a change in the observable behaviour which affects the available evidence so closing the feedback between cognitive conditions and empiric conditions; even if we consider the nexus between a change in expectations and a change in behaviour as deterministic, in the way economic theory routinely does, the empiric cause which connects a new observable event with an ensuing change in behaviour has to be conceived in principle as probabilistic since it is mediated in a crucial way by a probabilistic epistemic cause.

3   Weight of Argument and Second-order Epistemic Cause

As we have seen, the use of probabilistic causality in the *GT* is deeply rooted in the theory of probabilistic causality outlined in the *TP*, while its epistemological implications for an empirical field such as economics are clarified. Probabilistic causality is not at all the only specific link between *TP* and *TG*. A further crucial link is provided by another innovative concept introduced by Keynes in the *TP*: the *weight of argument*. Its emergence in the *TP* is determined by the inner conceptual logic of his approach while its pragmatic relevance remains in doubt;[5] in the *GT* Keynes provides important examples of its crucial relevance for an empirical field such as economics. This link has been already explored in the literature (see, e.g., Carabelli 1988; O'Donnel 1989; Runde 1990), while no one—to the best of my knowledge—has explored and discussed the link between the weight of argument and probabilistic causality neither in the *TP* nor in the *TG*. We intend to show that this link, only implicit in the Keynesian works, deserves

---

[5] In the *TP* Keynes explicitly confesses his doubts to the reader in a few passages, including the following: 'I do not feel sure that the theory of 'evidential weight' has much practical significance' (*TP*, p. 83).

to be made explicit and discussed. In order to do so we have to introduce briefly the very controversial concept of weight of argument. In the *TP* we may find at least three (apparently) different definitions of weight of argument. We argued elsewhere (Vercelli 1998) that the most satisfactory and comprehensive definition is the following:

> Given the argument $Q = (x \mid y)$, where x is a proposition (or set of propositions) that is true with a certain probability given the hypotheses and the background knowledge (i.e., the set of propositions y), the weight V of the argument Q is defined:
>
> $$V(Q) = K / (K + I)$$
>
> where K designates the relevant knowledge and I designates the relevant ignorance.

In intuitive terms the weight of argument measures the degree of completeness of the relevant knowledge of the epistemic subject. The range of values that the weight of an argument may assume goes from zero, when the epistemic subject believes to be in a state of complete ignorance in respect to the argument, to one when the epistemic subject believes to have a complete knowledge in respect to the argument. However, Keynes also clarifies that for an argument to be meaningful the relevant knowledge cannot be inferior to a certain minimum value ε (*TP*, p.78 ; see Fig.1).

Fig. 1

Now, we want to emphasise that also the causal arguments have a weight which expresses the reliability attributed to them by the epistemic subject. This is not made explicit in the *TP* by Keynes himself but descends from his conceptual framework since in his view all causal assertions in the epistemic sense are arguments connecting with some degree of probability a few propositions asserting the occurrence of one or more events, given a certain corpus of background knowledge, with the occurrence of one or more events which describe the effects. We may guess that Keynes did not make explicit this implication of his conceptual framework because having admitted his own doubts about the pragmatic relevance of the new concept of 'weight of argument', he did not perceive the potential relevance of its application to epistemic causality. However in the *GT* he is pushed by the inner logic of his economic analysis to attribute a crucial role to the weight of argument within a framework of probabilistic causal relationships between the relevant variables. A deep understanding of these passages of Keynes requires the introduction of a 'new' concept, new in the sense that it is only implicit in Keynes and never developed in the literature (to the best of my knowledge). I suggest to call it 'second-order cause' as it refers to a second-order measure of uncertainty, i.e., the weight of argument, the change of which may play a causal role, not only epistemic but also empiric. We may say in general that a change in the weight of a causal argument may change the decision strategy of a decision-maker who is led by it to adopt a different decision criterion. When the weight of argument is low it is rational to adopt a very prudential decision criterion such as the maximin criterion which is the most popular criterion adopted in case of full ignorance; when the weight of argument is high, i.e., the knowledge is almost complete, it is rational to exploit fully the available knowledge through a criterion such as that of the maximisation of expected utility (see Vercelli 1999). Of course the change in the decision criterion induced by a change in the weight of argument induces ceteris paribus a change in the behaviour of the decision-maker. Therefore also in this case an epistemic cause in principle may play the role of an empiric cause.

A proper use of second-order causality for epistemological and empirical analysis requires a rigorous definition. As a preliminary step in this direction, the following basic definition may be suggested:

C_i is a *pure* second-order epistemic cause of $B_{i'}$ in reference to the causal argument $Q = (B_{i'} | Z_i \cap A_i)$, iff:

(i) $P(A_i \cap Z_i) > 0$; $P(C_i \cap Z_i) > 0$,

(ii) $V(Q) > \varepsilon$, $V(Q | C_i) > \varepsilon$,

(iii) $P(B_{i'} | A_i \cap Z_i) = P(B_{i'} | A_i \cap Z_i \cap C_i)$,

(iv) $V(Q | C_i) \neq V(Q)$,

where the (i) and (ii) are conditions of meaningfulness. This definition characterises the event C_i a 'pure' second-order cause since its occurrence does not affect the first-order causal argument, as specified by the condition (iii) but only its weight, as specified by the condition (iv). In the real world the same event often, though not always, plays at the same time the role of first-order cause and of second-order cause but these two channels of transmission of a causal impulse must be kept separate because their effect may be quite different, even opposite (see the next section).

4 An Economic Example

We intend to clarify the concepts and the assertions introduced so far through a very simple economic example based on the theory of investment contained in the *GT* (see, e.g., Chick 1983). In order to determine the equilibrium quantity of investment of entrepreneurs in plants and machinery we have to equate demand and supply of investment. Under the usual assumptions which define a short-period equilibrium, the supply of new real capital S_k is given so that the variations of investment crucially depend on the variations of the demand D_k which is given by the sum of profits E_t expected from the new capital discounted at the rate of interest r:

$$(1) \qquad D_k = \Sigma E_t / (1 + r)^t, \qquad\qquad 1 \le t \le i.$$

The rate of interest that equates demand and supply of new real capital identifies the marginal efficiency of capital e_t. We may define the locus of possible equilibria through a curve which equates the market rate of interest r_t with the marginal efficiency of capital e_t. This curve is negatively sloped (see Fig. 2)[6] and may be shifted by a change in the expectations.

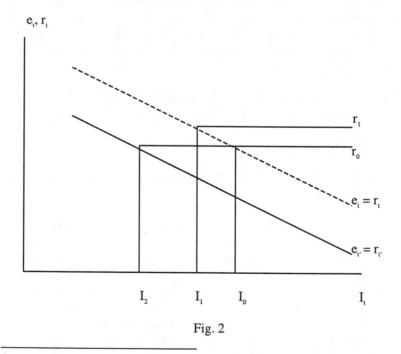

Fig. 2

---

[6] This is explained by neoclassical Keynesians as a consequence of the falling marginal productivity of capital, and by Keynes himself in terms of a limited stock of opportunities of investment ordered in terms of falling expected returns (see, e.g., Chick 1983; Vercelli 1991).

Therefore, since in the short term the supply curve of new real capital may be considered as constant, a change in investment may be caused only from the demand side, either by a change in the market rate of interest (empiric causality) or by a change in the expectations (first-order epistemic causality), or by a change in the weight of the arguments underlying the expectations and their change (second-order epistemic causality).

In the case of a change only in the observed market rate of interest (in Fig. 2, e.g., from r_0 to r_1), the curve of marginal efficiency of capital is not affected by this event so that the new value of the equilibrium investment is determined by the shift of the point where the exogenous rate of interest crosses the marginal efficiency of capital curve which determines a reduction of the investment from the original level I_0 to the new level I_1. This empiric causal relationship may be conceived as deterministic under the assumption, routinely adopted by economics, that the agent is rational in the sense that the option chosen maximises the expected returns. Let's assume, e.g., that ceteris paribus the market rate of interest increases. According to the relation (1) (represented in Fig. 2 by the downward-sloping curve of marginal efficiency of capital) we may say that

$$Q_0 = P\left[(\Delta I_t < 0) \mid Z_t \cap (\Delta r_t > 0)\right] = 1.$$

On the contrary, a change in the epistemic state of the decision-maker produces a shift in the marginal efficiency of capital curve which cannot be considered as deterministic. These shifts are based upon the probabilistic argument that connects the knowledge available to the decision-maker with the expected profits. In the most general perspective the basic argument may be expressed in the following way:

$$Q_1 = \phi(E_t \mid Z_t),$$

where $\phi(E_t \mid Z_t)$ represents the probability distribution of the expected profits conditional to the background information available to the decision-maker.

In our example a certain event is a first-order epistemic cause of the expected profits E_t whenever it modifies their probability distribution:

$$\phi(E_t \mid Z_t) \neq \phi(E_t \mid Z_t \cap C_t).$$

Let's assume that a central banker (say Greenspan) announces that he is not going to increase the rate of interest but warns the markets that the

Stock Exchange is characterised by a speculative bubble which may burst at any time. This new piece of information is likely to depress the expectations of the investors shifting downwards the marginal efficiency of capital curve from $e_t = r_t$ to $e_{t'} = r_{t'}$. Let's assume that C_t represents the new event (say the message of Greenspan), we have the following epistemic causal argument:

$$Q_2 = P\left[(\Delta E_t < 0) \,|\, Z_t \cap C_t\right] > P\left[(\Delta E_t < 0) \,|\, Z_t\right].$$

This *epistemic* causal argument easily translates into an *empiric* causal argument:

$$Q_3 = P\left[(\Delta I_t < 0) \,|\, Z_t \cap C_t\right] > P\left[(\Delta I_t < 0) \,|\, Z_t\right].$$

In principle, this empiric cause is independent of the empiric cause represented by the argument Q_0 and is probabilistic. This instantiates and confirms that also empirical causes are in principle probabilistic in the social field as soon as we consider the role played by the epistemic conditions in the causal chain. The two empiric causes represented by the argument Q_3, mediated by epistemic conditions, and the argument Q_0, independent of subjective epistemic conditions, do not exclude each other. As an example take the announcement by Greenspan of an increase in the rate of interest accompanied by a warning that the bubble of the stock exchange is likely to burst soon. In this case the two effects go in the same direction. However it is not necessarily so. Take the case of an announcement by Greenspan that he is going to reduce the rate of interest in order to react to a high degree of financial fragility. In this case the same announcement has two opposite effects: the reduction of the rate of interest should ceteris paribus increase the investment but its motivations (the fear of a generalised financial crisis) are likely to depress the expectations of the economic agents and to shift downwards the marginal efficiency of capital curve.

The analysis of the main causal determinants of investment would be gravely incomplete in the absence of a third causal factor of the utmost importance. The event C_t may reduce the weight of the epistemic argument underlying the epistemic causes involved in the process analysed either directly or indirectly. A warning by Greenspan that the expansion is about to end, whether or not accompanied by a change in the rate of interest or by other measures of policy, is likely to reduce the confidence of the agents over the future which may well be represented as a fall in the weight of the epistemic argument Q_1 underlying the existing expectations:

$$Q_4 = V(Q_1 \,|\, C_t) < V(Q_1)$$

This affects also the causal arguments Q_2 and Q_3 sharply increasing ceteris paribus the probability of a reduction in the investment. Therefore this second-order epistemic cause becomes an important independent empiric cause or an independent channel of transmission of a causal impulse. This example shows that the interplay of epistemic and empiric conditions in economics makes economic causality probabilistic, at least in principle, while the epistemic causes often play an indirect role of empiric causes. In addition we have seen that what we have called second-order epistemic cause may play a crucial role as empiric cause which confirms its operational role as well as that of the Keynesian concept of 'weight of argument'.

5 Concluding Remarks

The interplay of epistemic and empiric conditions of human behaviour is very complex and plays a crucial role in economic causality. We should therefore be suspicious of approaches, such as that of Granger causality, which oversimplify the issue. This approach is well rooted in, and fully consistent with, the prevailing approach of orthodox economics that assumes rational expectations; however the confusion between epistemic and empiric conditions of economic behaviour is typical of all this huge and influential literature, as is clearly revealed by the assumed identification of the subjective (or epistemic) probability distribution of the endogenous variables entertained by the economic agents with their 'objective' or empiric probability distribution.

In order to find a more satisfactory account of the interplay between epistemic and empiric conditions of the behaviour of economic agents we found useful to draw inspiration from the contributions of Keynes by linking together the *TP* with his successive economic contributions as developed in particular in the *GT*. As we have argued, the use of probabilistic causality in the *GT* is deeply rooted in the theory of probabilistic causality outlined in the *TP*, while its epistemological implications for economics are clarified in the *GT*. In particular, the awareness of the decision maker of her relevant ignorance, represented—and to some extent measured—by the weight of argument, plays the crucial role of ultimate stimulus of structural learning aiming to eliminate systematic mistakes, and determines the choice of the decision criterion on the part of a rational agent. This affects in a crucial way the economic behaviour of rational agents and so also the causal structure of economic events. Starting from these premises Keynes reaches a few important and innovative conclusions:

- Epistemic causality is in principle probabilistic in any field of science, since our knowledge is limited and is unable to detect deterministic links between events or variables even when we believe that such deterministic links exist in the real world. This is true not only of social sciences but also of natural sciences where Keynes accepts the then prevailing conviction in the deterministic nature of empiric causality.

- Since epistemic causality is probabilistic, also empiric causality is in general probabilistic in 'moral sciences' because in this case, differently than in natural sciences, the observed events are mediated in a crucial way by epistemic links, such as beliefs and expectations, which must be described in terms of probabilistic causality. This acquisition was extremely bald and innovative in the 1920s and 1930s when it was expressed by Keynes in his economic works.

- We have to distinguish different degrees of uncertainty which are expressed, and somehow measured, by the 'weight of argument'.

- Also the causal arguments have a weight. A change in this weight originated from the occurrence of an unexpected event represents what we have called here 'second-order epistemic causality' which often translates in an independent source of empiric causality which may be very important in practice, as illustrated in reference to Keynes's investment theory which has been briefly recalled and analysed here.

Though these important conceptual innovations play a crucial role in Keynes's economic works they have not been made clearly explicit by Keynes himself probably because he was mainly concerned with their practical implications. In any case they have been ignored in the subsequent literature, including most Keynesian economists, probably because, until very recently, it has been widely believed that there were not viable alternatives, at least of comparable rigour and analytical power, to the received decision theories which assume soft uncertainty and apply to a closed and stationary world. However in the last decade new decision theories have been worked out applicable to an open and evolutionary world characterised by hard uncertainty (important examples are Gilboa 1987, and Gilboa-Schmeidler 1989; an introductory survey may be found in Kelsey-Quiggin 1992, or Vercelli 1998). They are not less rigorous than the received theories and have been successfully applied to economic problems, including those studied by Keynes briefly and partially recalled here (see, e.g., Simonsen-Werlang 1991; Dow-Werlang 1992a, b). A new continent has been opened where the ideas of Keynes on probabilistic causality, which were much ahead of his time, may be fully developed.

References

Carabelli, A. (1988) *On Keynes's Method.* London: Macmillan.

Chick, V. (1983) *Macroeconomics after Keynes: A Reconsideration of the General Theory.* Oxford: Philip Allan.

Dow, J. and Werlang, S. R. C. (1992a) Excess volatility of stock prices and Knightian uncertainty. *European Economic Review,* 36, 631-638.

Dow, J. and Werlang, S. R. C. (1992b) Uncertainty aversion, risk aversion, and the optimal choice of portfolio. *Econometrica,* 60, 197-204.

Gilboa, I. (1987) Expected utility with purely subjective non-additive probabilities. *Journal of Mathematical Economics,* 16, 65-68.

Gilboa, I. and Schmeidler D. (1989) Maximin expected utility with a non-unique prior. *Journal of Mathematical Economics,* 18, 141-53.

Granger, C. W. J. (1969) Investigating causal relations by econometric models and cross-spectral methods. *Econometrica,* 37, 424-38.

Kelsey D., and Quiggin J. (1992) Theories of choice under ignorance and uncertainty. *Journal of Economic Surveys,* 6, 133-153.

Keynes, J. M. (1921/1973) *Treatise on Probability.* London: Macmillan. Reprinted in *CW* VIII: *The Collected Writings of John Maynard Keynes,* Vol. VIII, London: Macmillan.

Keynes, J. M. (1936/1973) *The General Theory of Employment, Interest and Money.* London: Macmillan. Reprinted in *CW* VII: *The Collected Writings of John Maynard Keynes,* Vol. VII, London: Macmillan.

Keynes, J. M. (1939) Professor Tinbergen's method. *Economic Journal,* 49, 34-51.

Keynes, J. M. (1973) *CW* XIV, *The General Theory and After, The Collected Writings of John Maynard Keynes,* vol XIV, London: Macmillan.

Keynes, J. M. (1979) *CW* XXIX, *The General Theory and After: A Supplement, The Collected Writings of John Maynard Keynes,* vol XXIX, London: Macmillan.

Morgenstern, O., and von Neumann, J. (1944) *Theory of Games and Economic Behaviour.* Princeton: Princeton University Press.

O'Donnel, R. (1989) *Keynes: Philosophy, Economics and Politics.* London: Macmillan.

Runde, J. (1990) Keynesian uncertainty and the weight of arguments. *Economics and Philosophy,* 6, 275-292.

Savage, L. J. (1956) *The Foundations of Statistics.* New York: John Wiley and Sons. Revised and enlarged edition, New York: Dover, 1972.

Simon, H. A. (1952) On the definition of the causal relation. *The Journal of Philosophy,* 49, 517-28.

Simonsen, M. H. and Werlang S. R. C. (1991) Subadditive probabilities and portfolio inertia. *Revista de Econometria,* 11, 1-19.

Suppes, P. (1970) *A Probabilistic Theory of Causality.* Amsterdam: North Holland.

Vercelli, A. (1989) Probabilistic causality and economic models: Suppes, Keynes and Granger. In K. Velupillai, (Ed.) *Nonlinear and Multisectoral Macrodynamics*. London: Macmillan.

Vercelli, A. (1991) *Methodological Foundations of Macroeconomics. Keynes and Lucas*. Cambridge: Cambridge University Press.

Vercelli, A. (1992) Probabilistic causality and economic models: A survey. In A. Vercelli and N. Dimitri (Eds.) *Macroeconomics: A Survey of Research Strategies*. Oxford: Oxford University Press.

Vercelli, A. (1998) Peso dell'argomento e decisioni economiche. In S. Marzetti Dall'Aste Brandolini and R. Scazzieri (Eds.), *La Probabilità in Keynes: Premesse e Influenze*. Bologna: CLUEB.

Vercelli, A. (1999) The recent advances in decision theory under uncertainty: A non-technical introduction. In L Luini, (Ed.), *Uncertain Decisions. Bridging Theory and Experiments*. Boston/Dordrecht/London: Kluwer, 237-260.

Wiener, N., (1958) The theory of prediction. In E. F. Beckenbach, (Ed.), *Modern Mathematics for Engineers, I*, chap. 8, New York: McGraw-Hill.

9

Bayesian Nets Are All There Is to Causal Dependence

WOLFGANG SPOHN
Universität Konstanz

1 Introduction

There are too many theories of causation to get into the focus of a small paper. But there are two in which I have a natural interest since they look almost the same: namely the theory of Clark Glymour, Peter Spirtes, and Richard Scheines, so vigorously developed since 1983[1] and most richly stated in Spirtes et al. (1993) (whence I shall refer to it as the SGS theory), and my own theory, published since 1978 in a somewhat irregular way. They look almost the same, but the underlying conceptions turn out to be quite dissimilar. Hence, the original idea for this paper was a modest one: simply to compare the philosophical basics of the two theories. However, no paper without a thesis! Therefore I have sharpened my comparison to the thesis written right into the title.

The plan of the paper is simple. Section 2 sets out the formal theory of Bayesian nets in an almost informal way, and Section 3 analyses the philosophical differences hidden in the common grounds. Section 4 briefly extends the comparison to the treatment of actions or interventions.

---

[1] The acknowledgments of Glymour et al. (1987) report that the work on that book took about four years.
Stochastic Causality.
Maria Carla Galavotti, Patrick Suppes and Domenico Costantini (eds.).
Copyright © 2001, CSLI Publications.

2   Causal Graphs and Bayesian Nets

Whenever we want to conduct a causal analysis in a given empirical field, we have to start by conceptually structuring this field. This is usually done by specifying a frame or a set U of variables characterizing the field. Each variable $A \in U$ can take some value from the set of its possible values. Thus, by specifying a value for each variable in U we specify some possible small world, some way how the empirical field characterized by the frame U may realize.

Variables should be conceived here as specific and not as generic variables. A generic variable would be something like social status or annual income which may take different values for different persons at different times. However, it is hard to find any causal order among generic variables. One then finds causal circles—high social status tends to generate high annual income, and vice versa—and one even finds apparent self-causation—social status tends to reproduce itself.

By contrast, a specific variable is something like my social status today or my annual income in 1998, not conceived as it actually is, which is given by some particular figure, but conceived as something which may take any value, say, between 0 and 1 billion Euros. There is a proper causal order among specific variables. For instance, there is no self-causation. If my social status today is high, it tends to be high tomorrow as well (though there is no guarantee, see the sudden fall of politicians), but this is a causal relation between two different specific variables.

Indeed, the causal structure within the frame U of specific variables is neatly captured by a causal graph over U which is nothing but a *DAG*, a *directed acyclic graph* <U,E> with U being its set of nodes and E being its set of edges. That the graph is directed means that its edges are directed, i.e. that E is an asymmetric relation over U, and that it is acyclic means that the directed edges don't form circles, i.e. that even the transitive closure of E is asymmetric.

Let me give a standard example (used by Pearl 1998 and elsewhere): U consists of five variables:

A_1:   season of a given year (spring, summer, fall, winter)

A_2:   rainfall during the season (yes, no)

A_3:   sprinkler during season (on, off)

A_4:   wet pavement (yes, no)

A_5:   slippery pavement (yes, no)

which we might plausibly arrange into the following DAG (if the variables refer to some place in Southern California):

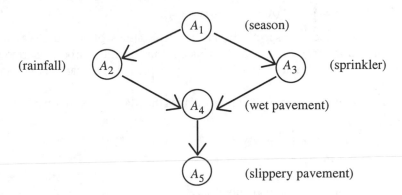

The DAG $<U,E>$ becomes a *causal graph*, if the edges in E are given a causal interpretation, i.e., if an edge $A \rightarrow B$ is interpreted as stating that A is directly influencing B, or that B is directly causally dependent on A, within the given frame U. Thus, so far the DAGs simply express the formal properties of direct causal dependence.

Specific variables have a specific temporal location. Hence, the variables in U are temporally ordered. So I shall add the natural constraint that in any edge $A \rightarrow B$ of a causal graph A temporally precedes B. Some philosophers oppose, but this is not the place to discuss their worries.

The next and crucial step is to introduce probabilities. The frame U generates, as mentioned, a space of possible small worlds the subsets of which may take probabilities according to some probability measure P. In particular, each event of the form $\{A = a\}$, stating that the variable A takes the value a, gets a probability. Accordingly, there is probabilistic dependence and independence among variables. More explicitly, we may define the sets X and $Y \subseteq U$ of variables to be *probabilistically independent given* or *conditional on* the set $Z \subseteq U$, i.e. $X \perp_p Y / Z$, iff for all x,y,z $P(X = x \mid Y = y, Z = z) = P(X = x \mid Z = z)$, i.e., iff, given any realization z of Z, any event about X is probabilistically independent of any event about Y.

Following SGS, we can state two conditions concerning a DAG $<U,E>$ and a measure P for U, in which $Pa(A)$ denotes the set of parents or immediate predecessors of the node A, $Nd(A)$ denotes the set of non-descendants of A, and $Pr(A)$ denotes the set of nodes temporally preceding A.

There is, first, the *Markov condition* (cf. Spirtes et al. 1993, pp. 53ff.) stating that for each $A \in U$ $A \perp_p Nd(A) / Pa(A)$. i.e., that each variable is independent from all its non-descendants given its parents. If the DAG agrees with the given temporal order this condition is equivalent to the apparently weaker condition that for each $A \in U$ $A \perp_P Pr(A) / Pa(A)$. This condition is also equivalent to the decomposability of P:

$$P(U = u) = \prod_{A \in U} P(A = a) \mid Pa(A) = x) ,$$

where a and x, respectively, are the realizations of A and $Pa(A)$ according to the realization u of U. This decomposability harbors enormous computational advantages so ingeniously exploited by Pearl (1988) and others.

For instance, the above example satisfies the Markov condition iff

$A_3 \perp_P A_2 / A_1$,
$A_4 \perp_P A_1 / \{A_2, A_3\}$, and
$A_5 \perp_P \{A_1, A_2, A_3\} / A_4$,

or iff, for all $a_1, ..., a_5$ realizing $A_1, ..., A_5$

$$P(a_1, ..., a_5) = P(a_1) \cdot P(a_2 \mid a_1) \cdot P(a_3 \mid a_1) \cdot P(a_4 \mid a_2, a_3) \cdot P(a_5 \mid a_4).$$

There is, second, the *minimality condition* (cf. Spirtes et al. 1993, pp. 53f.) stating that no proper subgraph of the DAG $<U,E>$ satisfies the Markov condition. Following Pearl (1988, p. 119) a DAG satisfying the Markov and the minimality condition is called a *Bayesian net(work)*. In a Bayesian net, the parents of a node thus form the smallest set of variables for which the relevant conditional independence holds.

For instance, the above example satisfies the minimality condition iff *none* of the following independencies holds:

$A_2 \perp_P A_1$,
$A_3 \perp_P A_1$,
$A_4 \perp_P A_2 / A_3$ and $A_4 \perp_P A_3 / A_2$, and
$A_5 \perp_P A_4$.

SGS further introduce a third condition, *the faithfulness condition* (cf. Spirtes et al. 1993, pp. 56), which is, however, more complicated and slightly less important so that I shall neglect it in the sequel.

So far, I have only introduced two distinct graph-theoretical representations: one of causal dependence between variables and one of conditional probabilistic dependence. However, the core observation of each probabilistic theory of causation is that there is a close connection between causal and probabilistic dependence, that the two representations indeed coincide, i.e., that *each causal graph is a Bayesian net*. Thereby, the Markov and the minimality condition turn into the *causal* Markov and the *causal* minimality condition. This means, to repeat, that the set of variables on which A di-

rectly causally depends within the frame U is *the* smallest set conditional on which A is probabilistically independent from all its other non-effects or, equivalently, from all other temporally preceding variables.[2] This assertion may indeed be used to *define* direct causal dependency within the frame U. At least I proposed to do so in Spohn (1976/8, section 3.3, in particular pp. 117f.). The definitional equivalence also follows from the assumptions made by SGS.

So far there is perfect agreement between SGS and me. However, there are also differences: first, concerning the development of causal theory, and second, concerning the understanding of the basic theory thus laid out. I shall dwell on the second point, but let me briefly mention the main differences of the first kind.

In my work, I did not use, and did not even think of, any graph-theoretical methods. These methods, graph-theoretic representations of independence relations, so-called d-separation, etc., were essentially introduced and pushed forward by Judea Pearl and his group after around 1985 (cf. Pearl 1988, pp. 132ff.). I am enthusiastic about these methods. They add powerfully to the strength, beauty, and vividness of the theory. Of course they are richly used by SGS. What I did have, however, in Spohn (1976/8, sections 3.2+3, with some variations translated in Spohn 1980), was the above-mentioned probabilistic definition of direct causal dependence and the full theory of conditional probabilistic independence on which this definition and the graph-theoretic methods rest, i.e., the graphoid and the semi-graphoid axioms, including the conjecture of their completeness (refuted by now) and the weaker conjecture of the completeness of the properties of direct causal dependence entailed by them (proved by now).[3]

Naturally, I wondered how the above account of causal dependence between variables may be founded on an account of causal relations between events or states of affairs or singular propositions. This is obviously philosophically important, but of little use in scientific and statistical methodology, and thus of no concern to SGS. The foundation seemed straightforward: the event $\{A = a\}$ is a *direct cause* of the event $\{B = b\}$ in the possible small world u if and only if both events occur in u, if $\{A = a\}$ precedes $\{B = b\}$, and if $\{A = a\}$ is positively relevant to $\{B = b\}$ according to P under the obtaining circumstances C, which are best identified with the event that all the variables preceding B (and differing from A, of course) take the values they take in u. Thus, the variable B directly causally depends on the

---

[2] That there is exactly one such set is a consequence of the properties of conditional probabilistic independence.

[3] For the conjectures see Spohn (1976/8, p. 105 and p. 119). For the positive and negative results see, e.g., the overview in Spohn (1994).

variable A iff some event about A is a direct cause of some event about B in some possible small world. For a long time, I was under the influence of the view of Suppes (1970, p. 58) that probabilistic causation cannot be transitive. In Spohn (1990) I changed my mind and started to prefer defining (direct or indirect) causation as the transitive closure of direct causation, though, as explained there, the issue is quite intricate.

Finally, in Spohn (1983, ch. 5 and 6; see also 1988) I have proposed the theory of ranking functions, as they are called nowadays, which yield a perfect deterministic analogue to probability theory, to conditional probabilistic dependence and independence, to the theory of Bayesian nets, and thus to the above account of probabilistic causation, and I have suggested there that this is how deterministic causation should be analyzed.[4]

So I have always moved within the philosophical confines. By contrast, Judea Pearl and his collaborators have done impressive work developing and utilizing the whole theoretical field for the purposes of artificial intelligence in a most detailed and fruitful way. And SGS have done impressive work developing sound statistical methodology on a sound philosophical basis, a different and in many respects much more difficult endeavor which starts to be successful in the big statistical community. Though all this work is addressed, to a large extent, to other departments, it contains a lot of high philosophical interest. But there is no place to further expand on this.

3   About the Causal Import of Bayesian Nets

Let me turn, then, to the interpretational differences between SGS and me which are my main concern. For this purpose, let us look again at the proposed definition: the variable A directly causally depends, within the frame U, on all and only the members of the smallest set of variables in U preceding A conditional on which A is probabilistically independent from all other variables in U preceding A. This definition hides two relativizations which deserve closer scrutiny.

First, direct causal dependence is obviously frame-relative according to this definition. The relativization would be acceptable, if it concerned only the direct/indirect distinction: what appears to be a direct causal dependency within a coarse-grained frame may well unfold into a longer causal chain within a more fine-grained frame. In this sense the frame-relativity is also accepted by SGS (cf. Spirtes et al. 1993, pp. 42f.). It's worse, however. The whole notion of causal dependence is frame-relative according to this definition: where there appears to be a direct or an indirect causal depend-

---

[4] A suggestion which I have coherently explained in English only in Spohn (2000).

ency within a coarse-grained frame, there may be none within a more fine-grained frame, and vice versa. This consequence seems harder to swallow.

The second relativization is better hidden. The talk of conditional independence refers, of course, to an underlying probability measure. Where does it come from?

It might come from reality, so to speak. This raises the question, of course, how to conceive of objective probabilities—a large question which I want to cut short by simply saying that they should best be understood as chances or propensities. This, however, is obscure enough. I have three reservations about using chances in the present context.

The first reservation is that chances are hard to find. But we want, and do, apply the probabilistic theory of causation almost everywhere, and in particular to fields where it is very unclear whether genuine chances exist. Almost all examples of SGS are from social sciences, medical sciences, etc. Maybe, if basic physics is chancy, everything else in the universe is chancy, too. But if so, we suffer from a complete lack of understanding of the chances, say, in economics or medicine, and whatever the probabilities are we are considering in these fields, they are certainly not suchlike chances.

A further reservation is that I find it very awkward in the meantime to talk of chancy events being caused (as has been most forcefully argued by Railton 1978). The idea behind genuine chances is that of partial determination without further determinability, and the idea behind causation is that of full determination. So, it's rather only the chances of events which are fully determined or caused and not the chancy events themselves. I certainly agree with Papineau (1989, pp. 308 and 320) that we need a probabilistic theory of causality in any case and that it is then largely a matter of terminology whether we should say that something that has raised the chance of an occurring event is among the causes of that event or only among the causes of the chance of that event. Still, my terminological preference is clear.

Mainly, however, my reservation is due to the fact that the above theory would be doomed as an analysis of causation if it starts with the notion of chance. The philosophical point of the enterprise is to elucidate the obscure notion of causal necessitation or full determination, and then the notion of chance or partial determination is presumably part of the package to be elucidated. To analyze the one in terms of the other does not seem helpful. I rather hoped to get a grip somehow on both notions together, on causation and chance.

If objective probabilities are thus to be avoided in the above definition of causal dependence, the only alternative is to use subjective probabilities. This is certainly an option, indeed the one I always preferred. However, it clearly amounts to a further relativization of causation to an epistemic subject or to its epistemic state. The above definition then says not what causal dependence *is*, but only how it is *conceived* by some epistemic subject.

This relativization is certainly in good Humean spirit. But even Hume who maintained it so bravely, was ambiguous and denied it at other places. Likewise, I have never been happy with these relativizations, but I did not get clear about how to get rid of them and what else to say about causation.

For instance, I could not see that the manipulability account of causation is of any help. Whether to explain the notion of something influencing something else by the notion of myself influencing something else or the other way around does not seem to make much of a difference. Moreover, actions, goals, etc. always deemed to me extraneous to the topic of causation. I found no help in the process theory of causation of Salmon (1984). Rich and illuminating as it is, its fundamental distinction between processes and pseudoprocesses leads in a large circle back to counterfactuals. So why not immediately engage into a counterfactual analysis of causation? Alluding to mechanisms is unhelpful since mechanisms seem to be nothing but suitably refined causal chains. The idea of energy transfer seems entirely beside the point when it comes to causation in the social sciences. Postulating a second-order universal of causal necessitation adds little in itself. And so forth.

So, the crucial question persisted: what else to say about causation? Only slowly it dawned upon me that I might, and indeed should, turn the inability to say more into a positive thesis. In a sense which I shall explain below there is nothing more to say about causation than I already did!

By contrast, these relativizations are plainly unacceptable to SGS, and this is, I admit, only common-sensical. They do not want, and do not pretend, to give an analysis of causation. They rather want to develop a theory over some undefined notion of causation, just as statistics is a big theory over some undefined notion of probability. So, in effect, they develop a theory jointly about causation and probability (cf. Spirtes et al. 1993, pp. 5ff. and 41ff.).

Their attitude, then, is this. Causal dependence, whatever it is, is ubiquitous. However, we are able to model only small parts of empirical reality by tentatively describing them by causal graphs and statistical hypotheses. The basic axiom of this model building is that these causal graphs are Bayesian nets, i.e., satisfy the Markov and the minimality condition introduced above (and also the faithfulness condition). The frame-relative definition of direct causal dependence is thus only an equivalence following from their axiom and has no explicative status. This shows clearly that their underlying conception is quite different from mine.

The natural follow-up question is: why should the axiom hold? SGS do not claim universal validity. The Einstein-Podolsky-Rosen paradox and quantum entanglement in general seem to provide a noticeable exception on which, however, I would like to be silent as well. But this does not diminish the success of the axiom elsewhere. They summarize their defense of the axiom in the following way:

> The basis for the Causal Markov Condition is, first, that it is necessarily true of populations of structurally alike pseudo-in-de-ter-ministic systems whose exogenous variables are distributed independently, and second, it is supported by almost all of our experience with systems that can be put through repetitive processes and whose fundamental propensities can be tested." (Spirtes et al. 1993, p. 64)

I am not quite satisfied by this. The first defense points to an interesting and important fact, but defers the issue to deterministic causation. And the second defense shows that we have a lot of intuitive skills and scientific knowledge in order to select appropriate sections of reality. But they continue the summary of their defense:

> Any persuasive case against the condition would have to exhibit macroscopic systems for which it fails and give some powerful reason why we should think the macroscopic natural and social systems for which we wish causal explanations also fail to satisfy the condition. It seems that no such case has been made.

Indeed, it is interesting how they argue about specific putative counter-examples. Their strategy is always the same: whenever there is a causal graph which is not a Bayesian net, there exists a suitable causal refinement of the original graph which is a Bayesian net. In the specific cases they discuss I find their argument convincing, for instance, when they reject the interactive forks of Salmon (1984, pp. 168ff.).[5]

But why should this strategy always work (with the disturbing exception already noticed)? Two possible explanations come to my mind. One possibility is that we have an *independent* notion of causation, and using that notion we generally happen to find suitable refined causal graphs which are Bayesian nets. But surely it is incredible that we merely happen to find these refinements. There should be a general reason for this success. Here one might continue in the following way:

> Basically, causation is deterministic, and then, given a specific conception of deterministic causation, we can specify very general conditions un-

---

[5] This rejection is of vital importance to their and my enterprise. If interactive forks were not only an apparently unavoidable, strange exception, as in the EPR paradox, but a perfectly normal and unsurprising phenomenon, as Nancy Cartwright argues again in this volume, then Bayesian nets would lose much of their interest, and my title thesis would simply be wrong.

der which such causal relationships get displayed in Bayesian nets. This is the strategy pursued by Papineau (1985). It is also the strategy behind SGS' theorem that (linear) pseudo-indeterministic systems, i.e., systems with a suitable (linear) deterministic extension in which the exogenous variables are independently distributed, satisfy at least the Markov condition (cf. Spirtes et al. 1993, pp. 58ff.).

This strategy is very illuminating as far as it goes. But I doubt that it works in the end. My reason for my doubt is that I don't believe that we have a workable theory of deterministic causation which could play this independent role. Rather I believe, as already indicated, that all our problems and arguments about probabilistic causation turn up all over again when deterministic causation is at issue.[6]

Hence, I don't think that the strategy presently envisaged works on the basis of deterministic causation. And I do not see any other independent notion of causation for which it has been, or could be, argued that it generally exhibits itself in Bayesian nets. So I am indeed skeptical of the whole approach.

How else might we explain that there always are suitably refined causal graphs which are Bayesian nets? The only other possibility which comes to my mind is to say that there is *no* independent notion of causation to be alluded to, that this *is* our understanding of causation. In other words: it is the structure of suitably refined Bayesian nets which decides about how the causal dependencies run. *We cannot regard B to be causally dependent on A unless we find a sequence of arrows or directed edges running from A to B in a suitably refined Bayesian net and unless, of course, this stays to be so in further refinements.* The last clause shows that the talk of suitable refinements is unnecessarily vague. In the final analysis it is the all-embracive Bayesian net representing the whole of reality which decides about how the causal dependencies actually are.

Of course, we are bound to have only a partial grasp of this all-embracive Bayesian net. Therefore it is important to have theorems telling under which conditions and to which extent our partial grasp is indicative of the final picture, that is, under which conditions the causal relations in a fine-grained Bayesian net are maintained in coarsenings. The theorem of SGS about pseudo-indeterministic systems is a good example. Clearly, however, the conditions to be specified in such theorems cannot be but assumptions about the shape of the final picture.

These remarks indicate how I propose to get rid of the two relativizations of causal dependence explained above. If the notion of causal dependence is prima facie frame-relative, we can eliminate this relativity only by

---

[6] See Spohn (2000) for some substantiation of this claim.

moving into the all-embracive frame containing all variables needed for a complete description of empirical reality. The all-embracive Bayesian net, then, does not distribute subjective probabilities over this frame in some arbitrary way. Rather, full information about the maximal frame should be accompanied by full information about the facts, so that subjective probabilities are optimally informed and thus objective at least in the sense proposed by Jeffrey (1965, ch. 12). In this way, the relativization of causal dependence to an epistemic state is eliminated as well.[7]

I am well aware that by referring to the all-embracive frame and to objective probabilities in this sense I am referring to entirely ill-defined and speculative entities. It is clear, moreover, that all causal theory can only deal with specific frames and specific Bayesian nets and their relations. Still, I find it philosophically inevitable to refer to such ill-defined entities, and the philosophical task is to try to strip them at least of some of their obscurity.

This finally explains my claim that in a sense there is no more to causal dependence than the above definition: this definition *with* its relativizations does all the theoretical work, and the move just proposed to eliminate these relativizations and thus to say what causal dependence really is is only a philosophical appendix adding no substantial theoretical content.

This needs two qualifying remarks. The first remark is that, even in the sense intended here, it is not wholly true that Bayesian nets exhaust all there is to the notion of causal dependence. I have hardly addressed the relation between time and causation and not at all the relation between space and causation, and both add considerably to the notion of causal dependence, i.e., to how the all-embracive Bayesian net has to look in the final analysis. By contrast, I have already expressed my doubts that such notions as action, mechanism, energy transfer, or process further enrich the notion of causal dependence. Anyway, whatever the further aspects of the notion of causal dependence, the theory of Bayesian nets covers its central conceptual content.

The second remark is that one must be very clear about the status of my claim that unrelativized, i.e., actual causal dependence is relativized causal dependence relative to the all-embracive frame and Jeffreyan objective probabilities. This is very much like the claim of Putnam (1980) that the ideal theory cannot be false. Both assertions are *a priori true*. Something is a priori true iff it cannot *turn out* to be otherwise. By contrast, something is necessarily true iff it cannot *be* otherwise. Hence, there is nothing metaphysically necessary about the truth of the ideal theory. The world could

---

[7] Or at least reduced. My vague formulations do not allow conclusions concerning the uniqueness of the objective probabilities thus understood.

easily be different from what the ideal theory says even given the truth of the ideally complete evidence on which it relies. But the world cannot turn out to be different from what the ideal theory says because this theory exhausts all factual and counterfactual means of evidence.

Similarly, causal dependence cannot turn out to be different from what it is in the all-embracive Bayesian net. But again this is only an epistemological claim, slightly more contentful than Putnam's claim, which has nothing to do with the metaphysics of causation. Indeed, I was completely silent on the latter. If I had wanted to say something about the metaphysics, I should have entered the whole of science, and then, of course, much more could be said.

Let me emphasize once more that I believe exactly the same story to apply to deterministic causation. There, again, Bayesian nets form the conceptual core of causal dependence, the only difference being that Bayesian nets are now constructed not in terms of probability measures, but in terms of ranking functions, their deterministic analogue.

4   Actions and Interventions

When I started to write about causation in Spohn (1976/1978), my real interest was decision theory. Therefore action variables were part of my picture from the outset. More precisely, I considered not only a set U of occurrence variables, as I called them for want of a better term, but also a set V of action variables. Thus the frame considered was always $U \cup V$. In decision contexts the task is to find the optimal action, action sequence, or strategy, and once one has found it, one starts executing it (unless weakness of will interferes). Hence, it does not make sense to assume the decision maker to have a probabilistic assessment of his *own* possible actions. For this reason I postulated that a decision model must not explicitly or implicitly contain any probabilities for the action variables in V (and thus took opposition to Jeffrey 1965).[8] So instead of considering one probability measure P over $U \cup V$ I followed Fishburn (1964, pp. 36ff.), and assumed a family $\{P_v\}$ of probability measures over U, parametrized by the possible action sequences v realizing V, which were to express probabilities of events over U conditional on v. It is straightforward then to extend the notion of conditional dependence and independence to such a family $\{P_v\}$, with the effect that relativized causal dependence can be explained relative to the frame $U \cup V$

---

[8] See also our exchange in Spohn (1977) and Jeffrey (1977). I still believe my principle "no probabilities for one's own options" to be correct and full of important consequences. It expresses, for instance, the most basic aspect of the freedom of the will since it exempts the will, i.e., willful actions, from causes, at least in the eyes of the agent. Cf. Spohn (1978, p 193).

in the way sketched above and that a causal graph over $U \cup V$ can be constructed which is a Bayesian net (in a slightly generalized sense). Consequently, all action variables are exogenous in that graph (but there may be more exogenous variables), and they introduce an asymmetry into the independence relation since occurrence variables can be (conditionally) independent from action variables, whereas the question whether an action variable is independent from another variable cannot arise simply because no probabilities are assigned to actions.[9]

A natural application of this account is Newcomb's problem, of course, which is basically a problem about the relation between probability and causality. As I observed in Spohn (1978, sect. 5.1), the account just sketched entails that among the four combinations of probabilistic dependence on and independence from action variables on the one hand and causal dependence on and independence from action variables on the other exactly one is impossible, namely the case that something is probabilistically dependent on, but causally independent from the action variables. But this, and only this, was the case Nozick (1969) worried about. Accordingly, there is no Newcomb problem, and two-boxing emerges as the only rational option. For more than 20 years I believed that this is the basic observation to be made about Newcomb's problem (though it is obvious, of course, that it does not exhaust the richness of the problem).[10]

When studying causation more closely later on, I neglected action variables for the sake of simplicity. But one can observe a growing interest in the explicit consideration of action variables in the theory of causation and the surrounding statistical and AI literature which certainly relates also to the triumph of the graph-theoretical methods. Thus, a theory of intervention or manipulation has become also a central part of the SGS theory.

Their picture is this (cf. Spirtes et al. 1993, pp. 75ff.). They start with an unmanipulated graph, as they call it, over a frame U. Then they consider one or several manipulations which they represent through a set V of variables enriching the original frame U in such a way that they are exogenous variables in the enriched or combined graph and directly manipulate or act on some variables in U. These intervention variables in V have a zero state which says: "Don't interfere!" or "Let it go!" If they take this state, the original unmanipulated graph stays in force. But if they take another state they enforce a new distribution on the directly manipulated variables irrespective, and thus breaking the force, of the ancestors of the directly ma-

---

[9] For all this see Spohn (1976/8, sections 3.1+2).

[10] Recently, however, I have converted to one-boxing. In Spohn (forthcoming) I expand the above consideration in a way which rationalizes one-boxing *within* the confines of causal decision theory.

nipulated variables in the unmanipulated graph. In the simplest case the new distribution will outright dictate a certain value to the directly manipulated variables. Their so-called manipulation theorem says then how to compute all the probabilities of the manipulated graph from the unmanipulated graph and the new distributions of the directly manipulated variables. All this provides also a nice and precise explanation of the epistemological difference between observing a variable to take a certain value and making it to take that value[11] which entail two quite different belief revisions (cf. also Meek, Glymour 1994, pp. 1007ff.).

However, the SGS theory of manipulation strikes me as being essentially equivalent with my old proposal just sketched. I did not distinguish a particular unmanipulated graph or, what comes to the same, a special zero state of the intervention variables, because there is not always a natural zero state—in the Newcomb situation you have to take one or two boxes, you cannot just let it go—and because non-interference or refraining seemed to me to be an action as well. One could, however, distinguish some values of action variables as such zero states in my framework and thus define the unmanipulated graph in the sense of SGS as the subgraph determined by these action variables taking their zero states. Their manipulation theorem then simply states the recursive decomposition of probabilities characteristic of Bayesian nets and their slight generalization to a probability family $\{P_v\}$.[12]

Again, a crucial difference lies in the fact that SGS build a very detailed statistical theory of prediction (of the effects of intervention) on their basic definitions.[13] Our basic agreement, however, is also displayed in our treatment of Newcomb's problem, where Meek, Glymour (1994, p. 1015) reach the same conclusion as the one I have sketched above.

To sum up: There is a large agreement between SGS and me in the formal basics of a probabilistic theory of causal dependence, including even the extension to actions or interventions. The main difference is that they

---

[11] A distinction which has been observed also by Kyburg (1980).

[12] The comparison extends to Pearl (1998, sect. 4).which summarizes his work on the role of actions in Bayesian networks. His procedure superficially differs from SGS's. Instead of expanding the original to a manipulated graph he includes action variables in the original graph (which, however, may merely be observed, from outside, as it were), and for representing actions as choices enforcing a certain value of the action variables he mutilates the original graph by cutting out all edges ending in actions variables. The mutilation also leads to a changed probability distribution, the same as the one described by SGS in their manipulation theorem. In Spohn (1978, sect. 5.2) I considered the very same problem—how to turn a theoretically detached view of a set of variables which does not give action variables a special role into a practically relevant view which does respect the special role of actions for the agent?—and I arrived at the very same cutting procedure.

[13] This remark applies *mutatis mutandis* to the work of Judea Pearl.

abstain from any bold statement about what causation is, wisely so for their purposes, whereas I have advanced and argued for the, positive or negative, thesis that from an epistemological point of view the theory of Bayesian nets exhaust, with the caveats mentioned, the theory of causal dependence.

References

Fishburn, P. C. (1964) *Decision and Value Theory*. New York: Wiley.

Glymour, C., Scheines, R., Spirtes, P. and Kelly, K. (1987) *Discovering Causal Structure*. New York: Academic Press.

Jeffrey, R. C. (1965/1983) *The Logic of Decision*. Chicago: Chicago University Press.

Jeffrey, R. C. (1977) A note on the kinematics of preference. *Erkenntnis,* 11, 135-141.

Kyburg, H. E. Jr. (1980) Acts and conditional probabilities. *Theory and Decision* 12, 149-171.

Meek, C. and Glymour, C. (1994) Conditioning and intervening. *British Journal for the Philosophy of Science* 45, 1001-1021.

Nozick, R. (1969) Newcomb's problem and two principles of choice. In N. Rescher et al. (Eds.), *Essays in Honor of Carl G. Hempel*. Dordrecht: Reidel, 114-146.

Papineau, D. (1985) Probabilities and causes. *Journal of Philosophy,* 82, 57-74.

Papineau, D. (1989) Pure, mixed, and spurious probabilities and their significance for a reductionist theory of causation. In P. Kitcher and W. C. Salmon (Eds.), *Minnesota Studies in the Philosophy of Science, vol. XIII, Scientific Explanation*, Minneapolis: University of Minnesota Press, 307-348.

Pearl, J. (1988) *Probabilistic Reasoning in Intelligent Systems: Networks of Plausible Inference*. San Mateo: Kaufmann.

Pearl, J. (1998) Graphical models for probabilistic and causal reasoning. In D. M. Gabbay and P. Smets (Eds.), *Handbook of Defeasible Reasoning and Uncertainty Management Systems, Vol. 1*. Dordrecht: Kluwer, 367-389.

Putnam, H. (1980) Models and reality. *Journal of Symbolic Logic* 45, 464-482.

Railton, P. (1978) A deductive-nomological model of probabilistic explanation. *Philosophy of Science,* 45, 206-226.

Salmon, W. C. (1984) *Scientific Explanation and the Causal Structure of the World*. Princeton: Princeton University Press.

Spirtes, P., Glymour, C. and Scheines, R. (1993) *Causation, Prediction, and Search*. Berlin: Springer.

Spohn, W. (1976/1978) *Grundlagen der Entscheidungstheorie*, Dissertation at the University of Munich 1976, published: Kronberg/Ts.: Scriptor, 1978 (references refer to the published version).

Spohn, W. (1977) Where Luce and Krantz do really generalize Savage's decision model. *Erkenntnis,* 11, 113-134.

Spohn, W. (1980) Stochastic independence, causal independence, and shieldability. *Journal of Philosophical Logic* 9, 73-99.

Spohn, W. (1983) *Eine Theorie der Kausalität*, unpublished Habilitationsschrift, University of Munich.

Spohn, W. (1988) Ordinal conditional functions. A dynamic theory of epistemic states. In W. L. Harper and B. Skyrms (Eds.), *Causation in Decision, Belief Change, and Statistics vol. II*. Dordrecht: Kluwer, 105-134.

Spohn, W. (1990) Direct and indirect causes. *Topoi* 9, 125-145.

Spohn, W. (1994) On the properties of conditional independence. In P. Humphreys (Ed.), *Patrick Suppes: Scientific Philosopher. Vol. 1: Probability and Probabilistic Causality*. Dordrecht: Kluwer, 173-194.

Spohn, W. (2000) Deterministic causation. In W. Spohn, M. Ledwig and M. Esfeld (Eds.), *Current Issues in Causation*. Paderborn: Mentis.

Spohn, W. (forthcoming) Dependency equilibria and the causal structure of decision and game situations. *Homo Oeconomicus*.

Suppes, P. (1970) *A Probabilistic Theory of Causality*. Amsterdam: North-Holland.

10

The Causal Interpretation of Regression Coefficients

PAUL W. HOLLAND

Educational Testing Service

1 Introduction and Notation

The causal interpretation of regression coefficients always involves a variety of assumptions that are plausible in some cases and not so easily assumed in others. I use the Neyman/Rubin model for causal effects to clarify these assumptions in a case of modest generality, i.e., a "prospective causal study" where the causal variable, x, is either a dummy variable indicating "treatment or control" or is a continuous measure of the "strength" of an exposure condition. In addition, I will assume that there are covariates available that can be used to justify the causal interpretation of the resulting regression coefficient of x through an assumption such as "strong ignorability".

There is widespread use of regression methods to assess the effects of causal factors such as experimental treatments or exposure to health hazards or to educational programs. Extensions of these regression methods to more complicated problems are also common and also strongly criticized (Freedman 1987, 1991, 1995, 1997). In this paper I will employ the formal model (or system of notation) introduced by Neyman (1923, 1935) for comparative experiments and extended to other types of non-randomized studies by Rubin (1974, 1978). This model is also used throughout the work of Robins (1985, 1997) in complex medical research applications. My purpose is to

Stochastic Causality.
Maria Carla Galavotti, Patrick Suppes and Domenico Costantini (eds.).
Copyright © 2001, CSLI Publications.

make the assumptions that underlie the "causal" use of regression methods in the easiest cases more explicit and possibly more understandable to a wider audience. Related work in the spirit of this paper can be found in Holland (1986, 1988, 1994), Holland and Rubin (1983, 1988), Rubin (1990), Rosenbaum (1984b), Robins (1986).

I make no effort here to justify attempts to draw causal inferences using regression-like statistical methods that involve *indirect* causal paths such as path analysis or instrumental variables. Rather, I will focus my attention on the cases where there is much less (but still some) controversy such as covariance adjustments and post-hoc matching in quasi-experiments (Cook and Campbell, 1979).

1.1 Prospective Causal Studies

The Neyman/Rubin notation is most easily understood in the context of a *prospective causal study* that has the general structure specified by this sequence of events.

1) "Subjects" or experimental "units" of study are identified.
2) "Baseline" or "pretest" information about these units is recorded.
3) The units are then either assigned to or "select themselves" to exposure to one of the "treatment" conditions or interventions of the study.
4) These units are then subsequently exposed to their assigned or self-selected treatment condition (and each unit is affected by this exposure in a manner that is unrelated to the exposure conditions of the other units).
5) At an appropriate later time an "outcome", "endpoint" or "post-test" measure is recorded for each unit in the study.

The type of study that this five-part schema is intended to cover includes all randomized comparative experiments as well as many types of pretest/post-test quasi-experiments or "observational studies". It should be emphasized that, properly interpreted, the Neyman/Rubin notation or model has application to other types of causal studies (see Holland (1988), Holland and Rubin (1988) and Robins (1997)), but, for my purpose here, prospective causal studies are already sufficiently complicated and inclusive. The condition mentioned parenthetically in 4), that the exposure conditions of the other units do not affect the outcomes associated with a given unit, is very important, and is clearly an assumption that would not be true in general. For example, in the study of infectious diseases your vaccination will affect my likelihood of contracting polio. I do not consider the restriction in 4) as a major problem with this approach to the study of causation because I regard causal inference as colored by the many specific circumstances of studies, of which the independent effect of the treatments on the units is only one of many.

1.2 The Neyman/Rubin Notation

The Neyman/Rubin notation is somewhat problematic because on the one hand it is easily confused with other more standard but less explicit notations and on the other because it involves sufficiently novel ideas it may appear to be unrelated to the practical work of applied scientists. Furthermore, the actual notations used by Neyman (1935) and by Rubin (1974, 1978), are both more complicated and more general that what I use here. My goal is simply to state the model in sufficient generality to give it utility in practical work without going to the extremes that are necessary in order to allow it to be applied to the most general causal studies and to account for the most minute details of such investigations.

The first point that should be emphasized is that the Neyman/Rubin notation was developed to fill in the gaps of the usual statistical notation so as to make certain hidden or tacit ideas more explicit. The second point is that it is neither the same as the usual notation nor is it far from it. I will try to warn the reader when things are different from what they might appear to be as well as indicate when standard notation is being used.

The prospective causal study begins with the "units", "subjects" or "cases" of the study, and I will denote the ith unit by the subscript "i". This is standard except that I will usually confuse sample and population because everything that I will discuss will be at the population level. This is just for convenience of discussion. It may help the reader to imagine that this is a discussion about a very large sample of units. I denote the population of units under study by P. Thankfully, for the most part P will lie quietly in the background without being noticed.

The baseline information that is collected or recorded for unit i will be denoted as a *vector* of numerical information, z_i. Again this is quite standard.

I will assume that there is a "causal" variable denoting a set of possible "treatments" or "exposure" conditions to which each unit in the study could be exposed. There are two cases of immediate interest for this discussion. In the first there are only two treatment conditions denoted by $x = 1$ (treatment) and $x = 0$ (control). In the second, a number, x, indicates a continuous measure of treatment "strength" or the value of a potentially large number of possible "graded" treatment-levels. An important aspect of causal variables designating such treatments-levels or intervention conditions is the assumption that the level of exposure for any unit *could have been different* from what it actually was. This condition excludes "attributes" of units (such as race, gender, age or pretest score) as causal variables in the sense that since such attributes "can not be other than what they are" they *can not* have "causal effects" in the sense that we will define in Section 2. This idea is discussed more extensively in Holland (1986 and 1988). This restriction on the meaning of causation is a problem for many who would like to talk about the causal effect of gender in the same way as they talk about the causal effect of an experimental treatment. Many experimentalists would regard such confusion as unproductive, as I do.

Each unit is exposed to one treatment level and the value of x to which i is exposed is denoted by x_i. This is also quite standard notation.

Finally we come to the outcomes or dependent variables in the study, and here is where a new notation is needed. We let $Y_i(x)$ denote the (numerical) response that would be recorded for unit i if unit i were exposed to treatment level, x. For each i, $Y_i(x)$ is a function of x. It should be emphasized that $Y_i(x)$ is not directly observed unless $x_i = x$. This is an important point because it is crucial to realize that the $\{Y_i(x)\}$ do not denote *observed data* like z_i and x_i do, but rather the $\{Y_i(x)\}$ are "potential outcomes" that lie behind the observed values of the outcome variable. For this reason, we denote the potential observations by capital letters to distinguish them from quantities that are directly observable, which we denote by lower case letters.

The connection between the potential outcomes and the actually observed outcomes is then given by the equation:

$$y_i = Y_i(x_i), \tag{1}$$

where y_i is the observed outcome or value of the dependent variable for unit i. The idea behind (1) is that to get from the potential outcomes $\{Y_i(x)\}$ to an observed outcome we must select the value of x in $Y_i(x)$ to be the value to which i is actually exposed, i.e., $x = x_i$, and then we obtain y_i from the potential observations, $\{Y_i(x)\}$, via (1).

The *observed data* for each unit i is the vector (z_i, x_i, y_i). The potential outcomes, $\{Y_i(x)\}$, are never observed for *all* values of x for a fixed unit, i, but only for the specific x-value to which i is actually exposed, x_i. It is sometimes said that the $\{Y_i(x)\}$ are "counterfactual" because they are not actual observations. I prefer to call them *potential observations* because they could have been observed had x_i been different than it was. I will mention what I regard as a better use of the term "counterfactual" in this setting in Section 3.

Up to this point I have not mentioned random variables or any of the usual fare of statistical models such as "errors of measurement". This is deliberate because I think such references obscure the basic ideas that are relevant. For me, the important aspect of units is their heterogeneity in the sense that the values $\{Y_i(x)\}$ can vary with i in all sorts of ways. I believe that the fruitful way to regard probability as entering the discussion of such models is through the sampling of units from the P rather than through viewing the units as having some sort of intrinsic probabilistic nature that produces the values of $\{Y_i(x)\}$ through some unknown (and probably unknowable) stochastic mechanism for each i. I am careful to define the $\{Y_i(x)\}$ as the values of Y "that would be *observed* for i if . . . ." and am silent as to how these values actually "come into existence". For a more extensive discussion of this view see Holland (1994).

However, I do wish to use the basic expected-value notation of random variables, *including conditioning* on the values of other variables. Hence, when I use a notation such as, $E(y)$, I mean the average value of y_i across the (large number of) units in P. Furthermore, by an expression such as,

$$E(y \mid x = x^*), \tag{2}$$

I will mean the average value of y_i across all of the (large number of) units in P for which $x_i = x^*$. This use of the expected-value notation can be justified from either a Frequentist or a Bayesian point of view, but this justification is not the main thrust of this paper so I will not go into it here. It should be noted, however, that within the expectation notation, the subscript i, denoting the unit, is suppressed because, within the scope of the $E(\ )$ operator, i is averaged over. This convention is perfectly consonant with the usual probability notation in which the notation for the elements of the sample space (in our case, P) is suppressed when using probabilities or expected-values, i.e., we use $P\{A\}$ rather than $P\{\omega: \omega$ in $A\}$, and $E(X)$ rather than $E(X(\omega))$, etc. In addition, because the subscript i is suppressed there can be confusion in a notation like that in (2) between x denoting the causal variable, x_i, and the value that x_i takes on. To avoid this confusion as much a possible I have adopted the convention of using asterisks, i.e., x^*, to denote the value that x_i takes on in the conditioning when using the expected value notation as exemplified in (2).

2   Using the Neyman/Rubin Notation

An important fact about the Neyman/Rubin model is the *Fundamental Problem of Causal Inference*, (Holland 1986), which is:

It is impossible in principle to *observe* $Y_i(x)$ for more than one value of x for any one unit, i.

Any procedure that claims to have avoided the *Fundamental Problem of Causal Inference* can always be shown to be based on untestable assumptions. Sometimes such assumptions are plausible, and sometimes they are not.

A basic definition that we are now in a position to make is that of a "unit-level causal effect". These are the differences

$Y_i(x) - Y_i(x^*)$ = the causal effect of x relative to x^* for unit i.

In the Neyman/Rubin notation, the unit-level causal effects are the *basic* quantities of interest in causal inference. However, the Fundamental Problem of Causal Inference is now immediately seen to be fundamental because it implies that unit-level causal effects are, themselves, *never directly observable*. Thus, we are always reduced to making assumptions that allow us to make some sort of conclusion or inference about these causal effects. This is the place where the Neyman/Rubin model might appear to be impractical for applied research, i.e., because its most basic parameters, the unit-level causal effects, are not directly observable. Furthermore, this is exactly the place where the potential exposability of all of the levels of x to any unit is seen to be crucial to the very foundations of the theory. The definition of causal effect requires this assumption so that the difference, $Y_i(x) - Y_i(x^*)$, is meaningful. It is the Fundamental Problem of Causal Inference and this definition of causal effect that makes causal inference both more interesting and more difficult than the simple computation of correlational and associational measures.

2.1 Causal Models:

A "causal model", in the sense that I will use this term here, is any assumption about the potential outcomes, $\{Y_i(x)\}$. There are many causal models in common use, some of these are discussed in Holland (1986) and (1988). For the present discussion I will limit my attention to "linear causal laws" (also called "linear statistical laws" by Pratt and Schlaifer (1984)) specified by the equation:

$$Y_i(x) = \alpha_i + \beta_i x. \tag{3}$$

A linear equation such as (3) might arise from some scientific law, such as Hooke's law for springs, or might arise as a linear approximation to a more complicated functional form for the dependence of $Y_i(x)$ on x. I will not distinguish between these two sources of the linear assumption, but will simply propose that from some point of view the linear form of the causal law should be scientifically plausible in the situations where the analysis in this paper is applicable. This is an extra-statistical assumption, and it might or might lead to testable consequences in real data. If the linear causal law is not plausible, then the account given here, of the causal interpretation of regression coefficients, must be made more complicated but is not essentially different.

It should be emphasized that equation (3) is *not a regression equation*; there is no error term! It is just the statement that for each unit in the study, the relationship between the "strength" of the causal variable or "treatment" and the potential outcomes is a straight line with intercept, α_i, and slope, β_i.

However, the slope coefficient, β_i, does have an interpretation in terms of unit-level causal effects. If x and x^* differ by one point on the scale of x, then we have

$$Y_i(x) - Y_i(x^*) \doteq \beta_i(x - (x-1)) = \beta_i. \tag{4}$$

Thus, the slope parameter, β_i, is an unit-level causal effect for any two values of x that differ by one x-point.

The relationship in (4) between unit-level causal effects and the slope parameter of a linear causal law is the key to the causal interpretation of regression coefficients given here. The problem to solve is to give conditions under which a regression coefficient for some variable will have a value closely related to β_i in (3) or (4).

It is now useful to introduce the notion of an **Average Causal Effect**, or ACE. In general, an ACE is any average of unit-level causal effects. The most general ACE has the form,

$$ACE = E(Y(x) - Y(x^*) | A), \tag{5}$$

where A denotes some collection of units defined in terms of either z_i or x_i or both. In (5) we again suppress the subscript i because it is being averaged over. As an example of an A in (5) we might use A = "all the units in the study", in which case the ACE is the average causal effect over all of **P**. But other cases might be of interest, for example A = "all units where i is male and for whom x_i = treatment group 1". In this case the ACE is for the males in treatment group 1.

If we now insert (4) into (5), we see that when a linear causal law can be assumed, the ACEs are just certain averages for the unit-level slope parameters, i.e.,

$$ACE = E(\beta | A) = \beta_A. \tag{6}$$

Hence, our interest in the causal interpretation of regression coefficients now turns to the question of when can a regression coefficient be identified as an ACE in the form of (6)?

For completeness, I mention here a simplifying "homogeneity" assumption. It is that β_i does not depend on the unit, i. In this case we have, $\beta_i = \beta_0$ for all units, i, and some value, β_0. In other words, for all i,

$$\beta_i = \beta_0 = E(\beta | A) \text{ for any set of units, } A. \tag{7}$$

Equation (7) is a special case of the assumption of "constant causal effect" discussed in Holland (1986, 1988). Under the "constant effect" assumption, the potential outcome, $Y_i(x) = \alpha_i + \beta_0 x$, can depend on i, but only through the intercept parameter, α_i. Under this assumption, the linear causal law is a series of parallel lines, all with slope, β_0. The constant effect assumption is very powerful and makes many analyses simple, because β_A in (6) no longer depends on the choice of A. I will not exploit this assumption in this paper other than to observe that in circumstances where it can be plausibly assumed in addition to the assumption of a linear causal law, any conclusion about ACEs also holds for the unit-level causal effects as well. These types of assumptions are how the unit-level causal effects can be brought into contact with observed data, thereby overcoming the Fundamental Problem of Causal Inference.

3 Making the Connection to Regression Models

Up to now, we have been content to define the basic structure of the data collection design as well as the causal connection between the potential outcomes and the causal variable x, i.e., equation (3). So far we have not identified the connection between any quantities that could be estimated with data and the causal parameters given by either the unit-level causal effects, β_i, or the average causal effects, $E(\beta \mid A) = \beta_A$. This leads us to the *"prima Facie* Average Causal Effects"*, or FACEs. The FACEs are what can be estimated using regression methods employing all of the observed data in hand, i.e., $\{(z_i, x_i, y_i)\}$.

Suppose we assume that the conditional expected-value function of y given x and z is linear, i.e., that

$$E(y \mid x = x^*, z = z^*) = a + bx^* + cz^*, \tag{8}$$

and that we use ordinary least squares to estimate the parameters of this function, i.e., a, b, and c. If the population conditional mean function has the linear form indicated in (8), then the coefficient, b, is of interest, because regression coefficients, such as b, are the main candidates for possible causal interpretations. (A more general discussion of (8) in terms of "best linear predictor" functions rather than conditional expectations is possible to construct, but goes beyond what I am attempting to do here.)

Applying the ideas behind equation (4) regarding causal effects, we are led to consider the difference between $E(y \mid x = x^*, z = z^*)$ and $E(y \mid x = x^{**}, z = z^*)$ for two values of x, (x^* and x^{**}) that differ by one point on the x-scale. The usual argument reveals that:

$$b = E(y \mid x = x^{**}, z = z^*) - E(y \mid x = x^*, z = z^*) \tag{9}$$

when $x^{**} = x^* + 1$, so I will say that b is a *prima Facie* Average Causal Effect, or FACE. By inserting the modifier "*prima facie*" I mean to indicate that a FACE is an ACE on the face of things but that its true nature needs to be further examined. Furthermore, FACEs are parameters that can be estimated with data directly, whereas ACEs are not so closely related to the observed data.

Thus, the problem before us is to identify assumptions that we must make in order for $b = \beta_A$ for some A, i.e., so that we may justify a *causal interpretation* of b as an ACE. The position taken in this paper, is that no other "causal interpretation" of b is really aligned with the causal structure of the problem as outlined earlier. That is, any *causal interpretation* of a regression coefficient that does not identify it as an ACE is incomplete, and, in the worst cases, entirely misleading.

To proceed, we need to bring the potential outcomes into the picture and this why they are so important. We use equation (1). Remembering to suppress the notation for the unit, i, in expectations, we have, for any choice of x^* and z^*,

$$E(Y(x^*)| x = x^*, z = z^*) = E(y| x = x^*, z = z^*) = a + bx^* + cz^*. \tag{10}$$

We can substitute y for $Y(x^*)$ in (10) because the conditioned value of x_i is x^*, so that $y_i = Y_i(x^*)$, in this case. The key idea behind this notation is that in (10) the value of x that modifies Y must be the same as that of the value of x_i in the conditioning in order for $Y_i(x)$ to refer to the observed value, y_i.

If, as before, $x^{**} = x^* + 1$, we now have

$$b = E(Y(x^{**})| x = x^{**}, z = z^*) - E(Y(x^*)| x = x^*, z = z^*)$$

$$= E(Y(x^{**}) - Y(x^*) + Y(x^*)| x = x^{**}, z = z^*)$$

$$\quad - E(Y(x^*)| x = x^*, z = z^*)$$

$$= E(Y(x^{**}) - Y(x^*)| x = x^{**}, z = z^*)$$

$$\quad + E(Y(x^*)| x = x^{**}, z = z^*) - E(Y(x^*)| x = x^*, z = z^*). \tag{11}$$

Next we replace $Y(x^{**}) - Y(x^*)$ by β_i using (4), and obtain

$$b = E(\beta| x = x^{**}, z = z^*)$$

$$\quad + E(Y(x^*)| x = x^{**}, z = z^*) - E(Y(x^*)| x = x^*, z = z^*), \tag{12}$$

which we may simplify to

$$b = \beta_A + \text{BIAS}, \tag{13}$$

where, using the notation of (6),

$$A = \{i: x_i = x^{**} \text{ and } z_i = z^*\} \tag{14}$$

and

$$\text{BIAS} = E(Y(x^*)| x = x^{**}, z = z^*) - E(Y(x^*)| x = x^*, z = z^*). \tag{15}$$

Thus, we have almost achieved our goal of expressing the FACE, b, in terms of an ACE, β_A, but we see there is a bias term defined in (15).

Our attention naturally turns to the term, BIAS. It is made up of two components, one that is "factual" and one that is "counterfactual". The factual term is

$$E(Y(x^*)|\, x = x^*, z = z^*) = E(y\,|\, x = x^*, z = z^*). \tag{16}$$

The quantity on the left-hand-side of (16) is "factual" because the right-hand-side of (16) is just the regression function evaluated at $x = x^*$, $z = z^*$. Thus, it is a quantity that, at least in principle, is estimable from the data collected in a prospective causal study.

The "counterfactual" term in (15) is

$$E(Y(x^*)|\, x = x^{**}, z = z^*), \tag{17}$$

and it is a quantity that is never *directly observable in principle* due to the Fundamental Problem of Causal Inference. In my opinion, this is a more serious use of the term "counterfactual" in the study of causation than generic references to the potential observations, $\{Y_i(x)\}$, because it pin-points exactly where the basic problems of causal inference actually lie, rather than just being a linguistic or philosophical nicety. In the first term of (15) we are asked for the average value of $Y_i(x^*)$ for the units who were exposed to $x_i = x^{**}$ rather than to x^*. There is nothing we can say about such a counterfactual term unless we make assumptions that are untestable with the data in hand, $\{(z_i, x_i, y_i)\}$. The quantity, BIAS, in (15) is a formula that specifies exactly where the problems of causal inference in non-randomized studies actually lie. Any attempt to deal with such studies in a serious way must ultimately address the BIAS term.

There is an important, commonly made, assumption for this purpose that we can make in special circumstances. It is called "strong ignorability given z" (Rosenbaum, 1984a, Rosenbaum and Rubin, 1984). This assumption asserts that, conditionally given z_i, x_i is *statistically independent* of $Y_i(x^*)$ over P for any x^* (this is one of the few places in this analysis where we need to actually make P explicit). When strong ignorability holds we have

$$E(Y(x^*)|\, x = x^{**}, z = z^*) = E(Y(x^*)|\, x = x^*, z = z^*) \tag{18}$$

and hence,

$$BIAS = 0. \tag{19}$$

If it holds, strong ignorability is just the ticket. In that case (13) then becomes

$$b = \beta_A. \tag{20}$$

This gives us a very general result. If (i) a linear causal law (i.e., (3)) holds, and if (ii) a *linear* regression function is correct (i.e., (8)) and (iii) if *strong ignorability* given z *can be assumed*, then the regression coefficient of x in (8) has a causal interpretation as an Average Causal Effect over a very special sub-population, i.e., A in (14). If these three assumptions are not plausible, then the coefficient of x in (8) is just another regression coefficient with nothing more going for it than its predictive power, which, of course, could be very useful in its own right.

3.1 Error Term Models:

My discussion would be incomplete if I did not make some reference to the use of "error terms" to formulate statistical models of relevance to causal inference. While I generally regard the use of models with error terms in them as one of the worst conceptual contributions of statistics to scientific study (Holland, 1995), they are so popular that some mention of their role needs to be mentioned. In particular, I will use error terms here to show a further analysis of the consequences of the strong ignorability assumption. My purpose is not to really use error term models, but to show how they are derivative of the more basic Neyman/Rubin model.

If we start with the potential observations and the linear causal law connecting them given in (3), we may express $Y_i(x)$ as

$$Y_i(x) = Y_i(0) + \beta_i x = \mu_0 + (Y_i(0) - \mu_0) + (\beta_i - \mu_\beta + \mu_\beta)x$$

or

$$Y_i(x) = \mu_0 + (Y_i(0) - \mu_0) + (\beta_i - \mu_\beta)x + \mu_\beta x, \tag{21}$$

where, $\mu_0 = E(Y(0))$, and $\mu_\beta = E(\beta)$.

Hence we have

$$Y_i(x) = \mu_0 + \varepsilon_i + \delta_i x + \mu_\beta x, \tag{22}$$

where,

$$\varepsilon_i = Y_i(0) - \mu_0, \text{ and } \delta_i = \beta_i - \mu_\beta. \tag{23}$$

Formula (22) is nothing more than a rewriting of the linear causal law in (3) and as such, it does not involve the observed data. If we substitute x_i for x, in it then we can introduce the observed data into the equation. This gives us:

$$y_i = Y_i(x_i) = \mu_0 + \varepsilon_i + \delta_i x_i + \mu_\beta x_i. \tag{24}$$

Now we can see that (24) is an error term model that relates the observed values of y_i and x_i together using the error terms, ε_i and δ_i. Error terms are supposed to have mean zero and both ε_i and δ_i automatically satisfy this condition when averaged over P.

In (24) we see two types of error terms that appear throughout the statistical literature. Neither one has anything to do with measurement error or even model error. Both quantities express the heterogeneity of the linear causal law in (3) across the units in P. Both ε_i and δ_i are functions of the potential observations, $\{Y_i(x)\}$, and thus are derivative of the "counterfactual aspects" of the Neyman/Rubin model of causal effects rather than being some sort of "stochastic shock" or "disturbance term" as they are so often described in the econometric literature. The $\{\varepsilon_i\}$ play the role of the typical "error term" about which various assumptions are often made such as "ε_i is independent of x_i". The nature of such assumptions is usually vague, but here it is a clearly an assumption about the statistical independence of $Y_i(0)$ and x_i over P. The $\{\delta_i\}$ play the role of "random effects" or "variance components" and reflect individual heterogeneity of the slopes in the linear causal law (3).

We may bring z into the picture by computing the conditional expectation in (8). It is

$$E(y \mid x = x^*, z = z^*) = E(\mu_0 + \varepsilon + \delta x + \mu_\beta x \mid x = x^*, z = z^*)$$

$$= \mu_0 + E(\varepsilon \mid x = x^*, z = z^*) + E(\delta x \mid x = x^*, z = z^*)$$

$$+ E(\mu_\beta x \mid x = x^*, z = z^*)$$

$$= \mu_0 + \mu_\beta x^* + x^* E(\delta \mid x = x^*, z = z^*) + E(\varepsilon \mid x = x^*, z = z^*). \qquad (25)$$

In (25) we see the potential of a non-linear regression function in (3) if the error terms do not have linear conditional expectations. Some simplification will occur if we now assume strong ignorability. Because δ and ε are just functions of the potential observations, strong ignorability given z will imply that x is independent of both of these error terms given z and hence we have:

$$E(\delta \mid x = x^*, z = z^*) = E(\delta \mid z = z^*) \qquad (26)$$

and

$$E(\varepsilon \mid x = x^*, z = z^*) = E(\varepsilon \mid z = z^*). \qquad (27)$$

Thus, the regression function in (3) becomes

$$E(y \mid x = x^*, z = z^*) = \mu_0 + \mu_\beta x^* + x^* E(\delta \mid z = z^*)$$

$$+ E(\varepsilon \mid z = z^*). \qquad (28)$$

Equation (28) indicates that the regression of y on x and z could be nonlinear in z and that z and x could interact multiplicatively. Such an equation shows the need to do careful data analysis in such situations in order to establish the nature of the dependence on z as well as the presence or the absence of the interactions between x and z. If x and z do not interact and if the dependence on z in (28), i.e., $E(\varepsilon \mid z = z^*)$ has been correctly estimated, then we see that the regression coefficient of x has a causal interpretation as an ACE, i.e., μ_β, the average of the β_i over all of P.

4 Discussion and Conclusions

Much more is possible to say and in this discussion section I will only hint at the possibilities. First of all, I hope this analysis makes it very clear that there are assumptions that might be plausible in certain circumstances that allow regression coefficients to have causal interpretations in a very strict but important sense, i.e., as ACEs. The problem is then to identify those cases where these assumptions are plausible and to disentangle them from the many cases where they are not. I think this is where judgement and relevant scientific knowledge enter the picture. Second, I have left open a wide door by focusing on the BIAS term in (15) It allows us to contemplate other types of assumptions (besides strong ignorability) about the causal structure of the problem, i.e., the $\{Y_i(x)\}$. What these new assumptions might be, I do not know in detail, but there are many possibilities that exploit the science of the situation and which could be used in particular circumstances. Indeed, I regard serious proposals of causal models, in the sense that I have use the term here, as one of the important tools in causal inference. This statement should not be construed to support the current meaning of the term, "causal modeling" that usually is much less serious than I propose here. Hooke's law is serious, $F = ma$ is serious, casual causal talk (including diagrams with many arrows) is rarely serious.

One situation in which strong ignorability is plausible arises when the covariate, z, strongly predicts the outcome, y. Examples of such strong prediction can arise in educational research because test performance is often an outcome or dependent variable in such studies and it is often possible to include similar "pretest" variables in z. It is often the case, with reliable tests at least, that pre and post tests are highly correlated. When the R^2 is large, there is not much of y left to predict, i.e., given z, y is more like a constant than a variable and constants are independent of any other variable. Such reasoning lies behind many appeals to large R^2's as a warrant to allow a causal interpretation of regression coefficients. The problem then becomes one of deciding which coefficients have causal interpretations? I hope that this discussion warns the reader that such a title ought not to be given to just any regression coefficient in a regression analysis with a large R^2.

When R^2 is not large, then, in my opinion, there is very little reason to hope that strong ignorability holds, unless there is some special knowledge available, such as that based on the design of the study. The easiest type of study to analyze, of course, is the completely randomized design, Holland (1986). More complicated examples include the "regression discontinuity design". In the latter case the value of x_i is completely determined by z_i, Rubin (1977), Cook and Campbell (1979).

References

Cook, T. D. and Campbell, D. T. (1979) *Quasi-experimentation: Design and Analysis Issues for Field Settings*. Boston: Houghton Mifflin.

Freedman, D. (1987) As others see us: a case study in path analysis. *Journal of Educational Statistics*, 12, 101-223.

Freedman, D. (1991) Statistical models and shoe leather. In P. Marsden, (Ed.), *Sociological Methodology*. Chapter 10, with discussion.

Freedman, D. (1995) Some issues in the foundations of statistics. *Foundations of Science*, 1, 19-83.

Freedman, D. (1997) From association to causation via regression. *Statistical Science*, 14, 243-258.

Holland, P. W. (1986) Statistics and causal inference. *Journal of the American Statistical Association*, 81, 945-970.

Holland, P. W. (1988) Causal inference, path analysis and recursive structural equations models. In C. Clogg, (Ed.), *Sociological Methodology*, 449-484.

Holland, P. W. (1994) Probabilistic causation without probability. In P. Humphreys (Ed.), *Patrick Suppes: Scientific Philosopher*, Vol. 1. Dordrecht: Kluwer, 257-292.

Holland, P. W. (1995) Some reflections of Freedman's critiques. *Foundations of Science*, 1, 41-67.

Holland, P. W. and Rubin, D. B. (1983) On Lord's Paradox. In H. Wainer and S. Messick (Eds.), *Principals* (sic) *of Modern Psychological Measurement*. Hillsdale NJ: Lawrence Erlbaum, 3-25.

Holland, P. W. and Rubin, D. B. (1988) Causal inference in retrospective studies. *Evaluation Review*, 12, 203-231.

Neyman, J. (1923) Sur les applications de la theorie des probabilites aux experiences agricoles: Essai des principes. *Roczniki Nauk Rolniczki*, 10, 1–51 in Polish: English translation by D. Dabrowska and T. Speed (1991) *Statistical Science*, 5, 463-480.

Neyman, J. (1935) Statistical problems in agricultural experimentation. *Supplement of the Journal of the Royal Statistical Society*, 2, 107-180.

Robins, J. M. (1985) A new theory of causality in observational survival studies—application of the healthy worker effect. *Biometrics*, 41, 311.

Robins, J. M. (1986) A new approach to causal inference in mortality studies with a sustained exposure period—application to control of the healthy worker survivor effect. *Mathematical Modelling*, 7, 1393-1512.

Robins, J. M. (1997) Causal inference from complex longitudinal data. In M. Ber-kane (Ed.) *Latent Variable Modeling with Applications to Causality.* New York: Springer-Verlag, 69-117.

Rosenbaum, P. R. (1984a) From association to causation in observational studies: The role of tests of strongly ignorable treatment assignment. *Journal of the American Statistical Association,* 79, 41-48.

Rosenbaum, P. R. (1984b) The consequences of adjustment for a concomitant vari-able that has been affected by the treatment. *Journal of the Royal Statistical Society, Series A,* 147, 656-666.

Rosenbaum, P. R. and Rubin, D. B. (1984) Estimating the effects caused by treat-ments. *Journal of the American Statistical Association,* 79, 26-28.

Rubin, D. B. (1974) Estimating causal effects of treatments in randomized and non-randomized studies. *Journal of Educational Psychology,* 66, 688-701.

Rubin, D. B. (1977) Assignment to treatment group on the basis of a covariate. *Journal of Educational Statistics,* 2, 1-26.

Rubin, D. B. (1978) Bayesian inference for casual effects: The role of randomiza-tion. *Annals of Statistics,* 6, 34-58.

Rubin, D. B. (1990) Formal models of statistical inference for causal effects. *Journal of Statistical Planning and Inference,* 25, 279-292.

11

Hierarchical Models and Partial Exchangeability

ATTILIO WEDLIN

University of Trieste

1 Introduction

In this paper I deal with a particular kind of stochastic dependence called partial exchangeability, which generalizes that of simple exchangeability. In particular we will point out that the usual statistical hierarchical models give rise to simple schemes of partial exchangeability.

It is known that the condition of exchangeability formalizes a situation of similarity or symmetry among observable variables in a way which is suitable for statistical applications. In fact, the model of independent and identically distributed (i.i.d.) variables can not be used when the statistician is willing to use some information on the values of some variables in order to make more accurate his probabilistic evaluations about the unknown elements of the model (estimation problem) or about the not yet observed variables (forecasting problem).

The condition of exchangeability typically implies a stochastic dependence among the observable variables, together with identical distributions of the variables. By definition, an infinite sequence of random variables is said to be exchangeable iff

$$(Y_1, Y_2, \ldots\ldots, Y_n) \overset{d}{=} (Y_{j1}, Y_{j2}, \ldots\ldots, Y_{jn})$$

for each positive integer n and every permutation $(j_1, j_2, \ldots\ldots, j_n)$ of the first n integer numbers.

Stochastic Causality.
Maria Carla Galavotti, Patrick Suppes and Domenico Costantini (eds.).
Copyright © 2001, CSLI Publications.

The notion of exchangeability was introduced by B. de Finetti in a lecture at the International Congress of Mathematicians held in Bologna in 1928 (De Finetti 1928). In spite of the presence at the Congress of outstanding probabilists such as M. Fréchet, P. Lévy, G. Darmois, S. Bernstein, A. Khinchin, E. Slutsky, R.A. Fisher, J. Neyman and others, the contribution of de Finetti and its implications for statistical inference went nearly unobserved. However, some years later, in 1935, de Finetti was invited in Paris by Fréchet in order to explain his viewpoint in probability and statistics.

In many situations a judgement of complete symmetry among the observable variables might be a too restrictive hypothesis, even though a judgement of partial symmetry might be supposed. A probabilistic scheme which may formalize such a condition is that known as *partial exchangeability*. As it is known, the notion of partial exchangeability was introduced in October 1937 by the Bruno de Finetti's paper "Sur la condition d'équivalence partielle" (De Finetti 1937); it was presented at the famous Colloque de Genève on the theory of probability. Let me recall the names of some participants at the meeting: M. Fréchet, P. Lévy, A. Wald, W. Feller, J. Neyman, G. Pòlya, W. Heisenberg, H. Cramér, H. Steinhaus, ... As in 1928 for the theory of exchangeable processes, the notion of partial exchangeability did not arouse much interest among the participants.

In 1938, de Finetti wrote a critical report (De Finetti 1938) about that meeting; summing up his own contribution, he said:

> The experience does not provide us with indications about probabilities, but facts which may exert an influence on our opinions or our probability assessments. And this in the sense that, conditional on the new circumstances that are now known, the prior probability was different from the prior unconditional probability so that the new opinion is not a denial of the previous one, but rather a necessary consequence ...
>
> In general, the probability of an event E after knowing the event A will be given by (Bayes theorem):
>
> $$P(E|A) = P(E \cap A) / P(A) = P(E) \cdot P(A|E) / P(A) .$$

The more common applications, as those which lead to probability assessment based on frequency observations, are only particular cases of this formula.

The simplest case is that of exchangeable events, that is events that can be permuted without modifying the probabilities; however some generalization is necessary because, in real situations, the various events hardly show a perfect analogy which justifies the assumption of exchangeability. More often, they will be partially exchangeable, that is we will have to distinguish various sub-classes and consider exchangeable only the events in the same subclass. In such a case we obtain mathematical formulas that are multidimensional versions of those relative to exchangeable events. Also the inferential reasoning takes place analogously: in assessing probabilities we will approach the observed frequency for events of the same kind (as for exchangeable events), but moreover there is an indirect influence depending on observed frequencies for events of other kinds.

A simple example to clarify these rather vague statements is the following: on tossing two imperfect coins which seem to be equal between them, we could initially be persuaded to assess a unique probability on the ground of the observed frequency in all tosses of both coins (without distinguishing the first group of tosses from the other). If the frequencies of the two coins are noticeably different between them when the number of tosses increases, then we will be willing to progressively distinguish the two probability assessments.

As regards *hierarchical statistical models*, a first suggestion about how to specify a probability distribution in several stages when the available information is scarce seems to be given by I. J. Good (1965) in his *The Estimation of Probabilities*. An early exposition of the hierarchical Bayesian approach was given by D. V. Lindley (1971) in the paper "The estimation of many parameters" and in the important seminal paper by D. V. Lindley and A. F. M. Smith (1972) "Bayes estimates for the linear model" was given a three stage hierarchical analysis for the normal linear model.

A simple first example of a hierarchical model is given by the so called "second-order probabilities": if E denotes an arbitrary set of events of interest and if our information on those events is rather vague so that we are unwilling to choose a unique probability on E, then we may consider a set P of admissible probabilities P and express our belief on those candidates by a second order probability Π so that our assessment about an event $E \in E$ is given by

$$\forall E \in E, \ \text{Prob}_\Pi (E) = \int_P P(E) \, d\Pi(P)$$

As a second example we show the parametric hierarchical linear model considered by Lindley and Smith in their above mentioned paper:

$$\mathbf{Y} / \Theta_1 = \theta_1 \sim N(A_1.\theta_1, C_1),$$
$$\Theta_1 / \Theta_2 = \theta_2 \sim N(A_2.\theta_2, C_2),$$

$$\Theta_2 / \Theta_3 = \theta_3 \sim N(A_3.\theta_3, C_3),$$

where $\mathbf{Y}$ is a random vector of observable variables, Θ_1 is a first level p_1-vector parameter, Θ_2 is a second level p_2-vector parameter and Θ_3 is a third level p_3-vector parameter ; usually, $p_1 \geq p_2 \geq p_3$. The matrices A_i and C_i (i=1,2,3) are assumed to be known.

In the Bayesian approach, the third level parameter Θ_3 is known so that the probability density $N(A_3.\theta_3, C_3)$ is completely specified; thus our overall prior opinion about the observable vector $\mathbf{Y}$ is expressed by

$$\mathbf{Y} \sim N(A_1.A_2.A_3.\theta_3, C_1 + A_1.C_2.A_1' + A_1.A_2.C_3.A_2'.A_1').$$

In the non-Bayesian approach, where all three vector parameters are considered as unknown constants, this probability density is formally interpreted as the likelihood function $l(\theta_3 / \mathbf{y})$; the likelihood functions for θ_2 and θ_1 may be obtained in terms of the densities: $f(y/\theta_2) \sim N(A_1.A_2.\theta_2, C_1 + A_1.C_2.A_1')$ and, respectively, $f(y/\theta_1) \sim N(A_1.\theta_1, C_1)$.

Here are some main reasons for using hierarchical models in parametric statistics:

a) to take account of the real condition of uncertainty in which the inferential procedures take place, that is of the usually limited prior information and the limited assessing-skills of the people and so on;

b) to combine information from related sources that we can not consider fully analogous, but only partially analogous;

c) to distinguish between structural, or objective, and subjective elements of the statistical model or of the prior information: usually, first level parameters constitute the structural element while hyperparameters constitute the subjective element of the model;

d) in the Bayesian framework, the decomposition of a prior joint distribution into several levels of conditional distributions may allow for an easier implementation of the statistical analysis using Monte Carlo simulation methods.

2   Something More about the Condition of Partial Exchangeability

As we said above, de Finetti introduced the scheme of partial exchangeability as a generalization of that of simple exchangeability in order to formalize some situations in which the observable events or variables are not quite analogous among them. There are several types of partial exchangeability, as is well illustrated in the paper by Diaconis and Freedman (1984), but we will consider only the most simple model of partial exchangeability which is constituted by a finite set of mutually correlated exchangeable processes.

An example of a real situation which we might formalize in terms of partial exchangeability is given by clinical responses to a drug used on several groups of patients, for instance male and female people. If we suppose that people in the same group are similar with reference to the effect of the drug, then we will assume for each group of observables the hypothesis of simple exchangeability. Finally, the opinion that the drug should produce the same type of effects on people belonging to different groups, although of different intensity, justifies the correlation between observables in different groups.

In formal terms, we will now present the definition of partial exchangeability and state the representation theorem for an infinite partially exchangeable process in the case of only two exchangeable subprocesses. The generalization to more than two subprocesses is straightforward.

Definition: the sequence of observable variables $\{Y_{ij} \mid i=1,2; j \geq 1\}$ is said to be a partially exchangeable process iff

$$(Y_{11},.....,Y_{1r} ; Y_{21},.......,Y_{2s}) \overset{d}{=} (Y_{1,j_1},.....,Y_{1,j_r} ; Y_{2,k_1},.....,Y_{2,ks})$$

for each pair of integers r,s and arbitrary permutations $(j_1,...........,j_r)$ and $(k_1,...........,k_s)$ of the first r and s integer numbers.

For infinite sequences of (simply) exchangeable random variables there is a famous result, known as de Finetti's representation theorem, according to which those random processes are representable as convex linear combinations of i.i.d. processes.

For infinite partially exchangeable sequences a similar theorem holds and a simple form of its statement is the following.

Representation theorem: for a given partially exchangeable process $\{Y_{ij} \mid i=1,2; j \geq 1\}$, there exist two real parameters θ_1, θ_2 and a joint distribution function $G(\theta_1,\theta_2)$ such that conditional on θ_1, θ_2 the observables Y_{ij} are all independent and have the same conditional distribution function $F_i(. / \theta_1,\theta_2)$ which depends on the first index i (i = 1,2) ; finally, we have the representation

$$F(y_{11},.....,y_{1r} ; y_{21},.......,y_{2s}) =$$

$$\iint [\prod_{j=1}^{r} F(y_{1j}/\theta_1,\theta_2)] \cdot [\prod_{k=1}^{s} F(y_{2k}/\theta_1,\theta_2)] \, dG(\theta_1,\theta_2) .$$

It can be proved that $G(\theta_1,\theta_2) \neq G_1(\theta_1) \cdot G_2(\theta_2) \Rightarrow \text{Cov}(Y_{1j},Y_{2k}) \neq 0$.

Thus such a theorem states that partially exchangeable infinite sequences are representable as convex linear combinations of processes whose variables are independent; moreover, they are supplied with the same univariate distribution inside the same subprocess.

A simple consequence of this fact is that the likelihood function of θ_1, θ_2 relative to the observed data $(y_{11},.....,y_{1r} ; y_{21},......,y_{2s})$ is given by the product

$$l(\theta_1, \theta_2 / y_{11}, ..., y_{1r} ; y_{21}, ..., y_{2s}) = \left[\prod_{h=1}^{r} l(\theta_1 / y_{1h}) \right] \left[\prod_{k=1}^{s} l(\theta_2 / y_{2k}) \right]$$

A first simple example of a partially exchangeable normal process is given by the following model:

$$Y_{ij} / \theta_1, \theta_2 \text{ independent r.v.'s }, \ j \geq 1, \ i = 1, 2;$$

$$f(y_{ij} / \theta_1, \theta_2) \ \sim \ N(\theta_i ; \sigma^2), \ \text{for each } j, \text{ with known } \sigma^2,$$

$$g(\theta_1, \theta_2) \ \sim \ N(\mu.1 ; \Gamma), \ \text{with known } \mu, \Gamma \text{ and } \gamma_{12} \neq 0.$$

It can easily be verified that $\{Y_{ij} \mid i = 1, 2; j \geq 1\}$ is a partially exchangeable normal process, which is characterized by the following moments:

$$E(Y_{ij}) \equiv \mu,$$

$$\text{Var}(Y_{ij}) = \sigma^2 + \gamma_{ii}, \ i = 1, 2,$$

$$\text{Cov}(Y_{ij}, Y_{ik}) = \gamma_{ii}, \ i = 1, 2,$$

$$\text{Cov}(Y_{1j}, Y_{2k}) = \gamma_{12}.$$

We will see later the statistical properties of this and other partially exchangeable observable processes.

3   A Two-stage Parametric Normal Hierarchical Model

A Bayesian hierarchical parametric (exchangeable) model is constituted by:

a)   a data generating process $\{Y_t ; t \geq 1\}$ which is assumed to be exchangeable,

b)   a common conditional density function $f(y_t / \theta)$ for all observable variables Y_t, where the unknown parameter θ belongs to a vector space of finite dimension,

c) a prior density function $g(\theta)$ for the unknown parameter which is decomposed into several conditional levels g_i $(\theta_{i-1} / \theta_i)$, i = 1,2,........,n , and a marginal density $g_n(\theta_n)$ so that it is

$$g(\theta) = \int \left[\prod_{i=1}^{n} g_i (\theta_{i-1} / \theta_i) \right] \cdot g_n (\theta_n) d\theta_1 .... d\theta_n \; ,$$

where $\theta_0 = \theta$.

More often, the number of the stages is two or three, although from a theoretical viewpoint it may be large as we wish. From a statistical viewpoint, the assumption of exchangeability is appropriate if we believe that, for any positive integer n, the observation of a finite sequence of n observable variables is as informative as that of any other sequence of n variables of the same process. From a mathematical viewpoint, the stochastic process $\{Y_t ; t \geq 1\}$ is exchangeable if and only if any finite joint distribution function is symmetric, that is invariant with respect to the order of its arguments. The above mentioned de Finetti's representation theorem for (simply) exchangeable processes supplies us with the integral representation

$$F(y_1,........,y_n) = \int_{\{\theta\}} \left[\prod_{i=1}^{n} F(y_i / \theta) \right] \cdot g(\theta) d\theta \; ,$$

which is a particular case of the representation for partially exchangeable processes.

Let us now consider a simple example of a two-stage exchangeable hierarchical model:

$$Y_t / \theta, \lambda \; \sim \; N(\theta; \sigma^2), \quad t \geq 1, \; \sigma^2 \text{ known},$$

$$\Theta / \lambda \; \sim \; N(\lambda; s^2), \quad s^2 \text{ known},$$

$$\Lambda \; \sim \; N(\mu_0; \sigma_0^2), \quad \text{completely specified}.$$

It is easy to realize that these assumptions imply the following marginal density for Θ

$$\Theta \; \sim \; N(\mu_0; \sigma_0^2 + s^2),$$

and the following joint density for (Y_t, Θ, Λ)

$$\begin{bmatrix} Y_t \\ \Theta \\ \Lambda \end{bmatrix} \sim N^{(3)} \left(\mu_0 \cdot 1; \begin{bmatrix} \sigma^2 + \sigma_0^2 + s^2 & \sigma_0^2 + s^2 & \sigma_0^2 \\ \sigma_0^2 + s^2 & \sigma_0^2 + s^2 & \sigma_0^2 \\ \sigma_0^2 & \sigma_0^2 & \sigma_0^2 \end{bmatrix} \right) .$$

It is also easy to verify that the data generating process $\{Y_t ; t \geq 1\}$ is an exchangeable normal process which is characterized by the common mean value μ_0, the common variance $\sigma^2 + \sigma_0^2 + s^2$ and the common covariance $\sigma_0^2 + s^2$ for all variables Y_t and pairs of variables Y_s, Y_t.

If we suppose to know the value y of the observable variable Y_t, we are able to make inference on the first level parameter Θ and the second level parameter Λ; the posterior densities are as follows:

$$\Theta / Y_t = y \sim N \{ [\sigma^2/(\sigma^2 + \sigma_0^2 + s^2)].\mu_0 + [(\sigma_0^2 + s^2)/(\sigma^2 + \sigma_0^2 + s^2)].y ;$$

$$[\sigma^2(\sigma_0^2 + s^2)/(\sigma^2 + \sigma_0^2 + s^2)]\} ,$$

$$\Lambda / Y_t = y \sim N \{ [(\sigma^2+s^2)/(\sigma^2 + \sigma_0^2 + s^2)].\mu_0 + [\sigma_0^2/(\sigma^2 + \sigma_0^2 + s^2)].y ;$$

$$[\sigma_0^2(\sigma^2 + s^2)/(\sigma^2 + \sigma_0^2 + s^2)] \} .$$

Let us now suppose that the data consist of two sequences of observable variables $\{Y_{1t} ; t \geq 1\}$ and $\{Y_{2t} ; t \geq 1\}$, each of which is assumed to be exchangeable. We could adapt the preceding normal model to our new situation by assuming that the first level parameter is a vector parameter $\Theta = (\Theta_1, \Theta_2)'$ and choosing

$$Y_{it} / \theta, \lambda \sim N(\theta_i ; \sigma^2), \quad t \geq 1, \quad i = 1, 2; \quad \sigma^2 \text{ known};$$

$$\Theta / \lambda \sim N^{(2)}(\lambda.1 ; s^2.I_2), \quad s^2 \text{ known};$$

$$\Lambda \sim N(\mu ; \sigma_0^2), \quad \text{completely specified};$$

finally, we assume that conditionally on all events $(\Theta_1 = \theta_1 \cap \Theta_2 = \theta_2 \cap \Lambda = \lambda)$, all observable variables Y_{it} are independent among them.

From the second and third assumptions we may obtain the following joint density for all parameters:

$$\begin{bmatrix} \Theta_1 \\ \Theta_2 \\ \Lambda \end{bmatrix} \sim N(\begin{bmatrix} \mu.1 \\ \mu \end{bmatrix}; \begin{bmatrix} \sigma_0^2.1.1'+s^2.I_2 & \sigma_0^2.1 \\ \sigma_0^2.1' & \sigma_0^2 \end{bmatrix}) ,$$

from which we realize that both parameters Θ_1 and Θ_2 have a density function $N(\mu, s^2 + \sigma_0^2)$ and $\text{Cov}(\Theta_1, \Theta_2) = \sigma_0^2$. The marginal joint density for Θ_1 and Θ_2, of the type $N^{(2)}(\mu.1; \sigma_0^2 11'+s^2 I_2)$, is therefore a particular case of the distribution $N(\mu.1 ; \Gamma)$ in the example at page 4, with $\gamma_{12} = \sigma_0^2$. Thus, it can easily be proved that each process $\{Y_{it} ; t \geq 1\}$, $i = 1,2$, is an exchangeable normal process which is characterized by mean values μ, variances $\sigma^2 + s^2 + \sigma_0^2$ and covariances $s^2 + \sigma_0^2$. We may also verify that Cov $(Y_{1r}, Y_{2t}) = \sigma_0^2$, for all choices of positive integer r and t, so that this pair of mutually correlated exchangeable processes constitutes a simple case of a partially exchangeable process, according to the example at page 190.

We will now examine the statistical implications on all unknown parameters of such a probabilistic structure. We will assume we can observe m variables of the first process $\{Y_{1t} ; t \geq 1\}$ and n variables of the second one $\{Y_{2t} ; t \geq 1\}$ so that our statistical observations are given by $(y_{11}, y_{12}, \ldots, y_{1m} ; y_{21}, y_{22}, \ldots, y_{2n}) = (\mathbf{y_1}', \mathbf{y_2}')$. We will determine the corresponding posterior distributions for the first level parameters Θ_i (i = 1, 2) and that for the second level parameter Λ.

If $A_1 = \text{Cov} (\mathbf{Y_1})$ denotes the covariance matrix of the observable variables $Y_{11}, Y_{12}, \ldots, Y_{1m}$, $A_2 = \text{Cov} (\mathbf{Y_2})$ that of $Y_{21}, Y_{22}, \ldots, Y_{2n}$, S_{12} a (m,n) matrix whose elements are all equal to σ_0^2, S_{21} a similar (n,m) matrix, $\mathbf{1_r}$ an r- vector with all elements equal to 1, then we can specify the joint density of $(\mathbf{Y_1}', \mathbf{Y_2}', \Theta_1, \Theta_2, \Lambda)'$ as a normal density with mean vector given by $\mu.\mathbf{1}_{-m+n+3}$ and covariance matrix :

$$
\begin{bmatrix}
A_1 & S_{12} & (s^2 + \sigma_o^2)1_m & \sigma_0^2 1_m & \sigma_0^2 1_m \\
S_{21} & A_2 & \sigma_0^2 1_n & (s^2 + \sigma_0^2)1_n & \sigma_0^2 1_n \\
(s^2 + \sigma_0^2)1'_m & \sigma_0^2 1'_n & s^2 + \sigma_0^2 & \sigma_0^2 & \sigma_0^2 \\
\sigma_0^2 1'_m & (s^2 + \sigma_0^2)1'_n & \sigma_0^2 & s^2 + \sigma_0^2 & \sigma_0^2 \\
\sigma_0^2 1'_m & \sigma_0^2 1'_n & \sigma_0^2 & \sigma_0^2 & \sigma_0^2
\end{bmatrix} .
$$

It follows that the posterior distributions of one first level parameter, say $\Theta_1 / \mathbf{y_1}, \mathbf{y_2}$, and of the second level parameter $\Lambda / \mathbf{y_1}, \mathbf{y_2}$ are all of normal type with parameters given by :

$$
E(\Theta_1 / \mathbf{y_1}, \mathbf{y_2}) = \mu + [(s^2 + \sigma_0^2).\mathbf{1'}_m , \sigma_0^2.\mathbf{1'}_n] .D^{-1}. \begin{bmatrix} y_1 - \mu.1_m \\ y_2 - \mu.1_n \end{bmatrix} ,
$$

where $D = \begin{bmatrix} A_1 & S_{12} \\ S_{21} & A_2 \end{bmatrix}$,

$$
V(\Theta_1 / \mathbf{y_1}, \mathbf{y_2}) = (s^2 + \sigma_0^2) - [(s^2 + \sigma_0^2).\mathbf{1'}_m , \sigma_0^2.\mathbf{1'}_n] .D^{-1}.
$$
$$
[(s^2 + \sigma_0^2).\mathbf{1'}_m , \sigma_0^2.\mathbf{1'}_n]' ;
$$

$$E\left(\Lambda / \mathbf{y}_1, \mathbf{y}_2\right) = \mu + \left(\sigma_0^2 \, .\mathbf{1'}_{m+n}\right).D^{-1}. \begin{bmatrix} y_1 - \mu.\mathbf{1}_m \\ y_2 - \mu.\mathbf{1}_n \end{bmatrix},$$

$$V\left(\Lambda / \mathbf{y}_1, \mathbf{y}_2\right) = \sigma_0^2 - \left(\sigma_0^2 \, .\mathbf{1'}_{m+n}\right).D^{-1}.\left(\sigma_0^2 \, .\mathbf{1'}_{m+n}\right)'.$$

At this point, it is worth making some remarks: the inference on each first-level parameter, say Θ_1, utilizes both vectors of observations, but in a different way, as we can see straightforwardly from the above expressions. As a limit-case, such an inference might take place also if we are only able to observe variables whose distributions do not depend on Θ_1, that is the observables Y_{2t}, $t \geq 1$. The reason of this "cross-inference" is the correlation between the different data-generating processes, which is determined by the correlation between Θ_1 and Θ_1.

4   Combining Information from Various Sources of Data

The prototype of inferential problems in conditions of simple exchangeability is as follows: if a coin lands heads up, what inference may be drawn about the probability that the same coin still lands heads up in the next flip?

In situations of partial exchangeability, we might instead consider problems as the following: given that coin A lands heads up, what inference may be drawn about the probability that coin B lands heads up in the next flip? More generally, given that we have obtained h and, respectively, k heads in m and n flips of two coins, A and B, what conclusion may be drawn about the probability that coin B lands heads up in the next flip? Obviously, a preliminary subjective assumption of similarity of the two coins A and B is a necessary condition for giving some meaning to this question.

This kind of problems belongs to the field of statistical meta-analysis, where typically results of several experiments on the same phenomenon, carried out in different places with different subjects, are pooled together in order to improve the quality of inferences. In this realm, a rather simple framework is constituted by the condition of partial exchangeability

For the sake of illustration, we will present a simple partially exchangeable statistical model which allows the combination of information from two different but correlated sources of data. It is obtained by slightly modifying the preceding model with regard to the known second moments:

$$Y_{it} / \theta, \lambda \sim N(\theta_i ; \sigma_i^2), \quad t \geq 1, \ \sigma_i^2 \text{ known}, \ i = 1, 2;$$

$$\Theta / \lambda \sim N^{(2)}\{ \lambda.\mathbf{1}_2 ; \ \text{diag}(s_1^2, s_2^2) \}, \quad s_i^2 \text{ known}, \ i = 1, 2;$$

$$\Lambda \sim N(\mu_0 ; \sigma_0^2), \quad \mu_0, \sigma_0^2 \text{ known}.$$

In this case, we obtain the following joint density function for all unknown parameters:

$$\begin{bmatrix} \Theta_1 \\ \Theta_2 \\ \Lambda \end{bmatrix} \sim N(\begin{bmatrix} \mu_0 \cdot 1 \\ \mu_0 \end{bmatrix}; \begin{bmatrix} diag(s_1^2, s_2^2) + \sigma_0^2 1.1' & \sigma_0^2 \cdot 1 \\ \sigma_0^2 \cdot 1' & \sigma_0^2 \end{bmatrix}) \,,$$

which implies that each observable process $\{Y_{ij} | j \geq 1\}$, i=1,2 , is an exchangeable normal process with mean values equal to μ_0, variances $\sigma_i^2 + s_i^2 + \sigma_0^2$ and covariances $s_i^2 + \sigma_0^2$; finally, we have $Cov(Y_{1j}, Y_{2k}) = \sigma_0^2$.

The joint density function of the vector $(\mathbf{Y}_1', \mathbf{Y}_2', \Theta_1, \Theta_2, \Lambda)'$ is still normal, with mean vector given by $\mu_0 \cdot 1 \underset{-m+n+3}{}$, and covariance matrix:

$$\begin{bmatrix} A_1 & S_{12} & (s_1^2 + \sigma_o^2)1_m & \sigma_0^2 1_m & \sigma_0^2 1_m \\ S_{21} & A_2 & \sigma_0^2 1_n & (s_2^2 + \sigma_0^2)1_n & \sigma_0^2 1_n \\ (s_1^2 + \sigma_0^2)1'_m & \sigma_0^2 1'_n & s_1^2 + \sigma_0^2 & \sigma_0^2 & \sigma_0^2 \\ \sigma_0^2 1'_m & (s_2^2 + \sigma_0^2)1'_n & \sigma_0^2 & s_2^2 + \sigma_0^2 & \sigma_0^2 \\ \sigma_0^2 1'_m & \sigma_0^2 1'_n & \sigma_0^2 & \sigma_0^2 & \sigma_0^2 \end{bmatrix}$$

In order to illustrate the so called cross-inference situation, we will now confront the posterior distributions of one first level parameter, say Θ_1 , connected with observing only $\mathbf{Y}_2$ or both $\mathbf{Y}_1, \mathbf{Y}_2$. We obtain the following posterior densities:

$$\Theta_1 / \mathbf{y}_2 \sim N[\mu_0 + \sigma_0^2 \cdot 1'_n \cdot A_2^{-1} \cdot (\mathbf{y}_2 - \mu_0 \cdot 1'_n) ; (s_1^2 + \sigma_0^2) - \sigma_0^2 \cdot 1'_n$$
$$\cdot A_2^{-1} \cdot \sigma_0^2 \cdot \mathbf{1}_n] \,,$$

$$\Theta_1 / \mathbf{y}_1, \mathbf{y}_2 \sim N \{ \mu_0 + [(s_1^2 + \sigma_0^2) \cdot 1'_m , \sigma_0^2 \cdot 1'_n] \cdot D^{-1} \cdot \begin{bmatrix} y_1 - \mu_0 \cdot 1_m \\ y_2 - \mu_0 \cdot 1_n \end{bmatrix} ;$$

$$(s_1^2 + \sigma_0^2) - [(s_1^2 + \sigma_0^2) \cdot \mathbf{1}'_m , \sigma_0^2 \cdot 1'_n] \cdot D^{-1} \cdot [(s_1^2 + \sigma_0^2) \cdot \mathbf{1}'_m , \sigma_0^2 \cdot 1'_n]' \} \,,$$

where D denotes the covariance matrix of the vector $(\mathbf{Y}_1', \mathbf{Y}_2')'$ in this model.

We can see that the influence on the posterior density of Θ_1 of observed vectors $\mathbf{y}_1$, $\mathbf{y}_2$ is different in relation to the values of the covariances $Cov(\Theta_1, Y_{1t}) = s_1^2 + \sigma_0^2$ and $Cov(\Theta_1, Y_{2t}) = \sigma_0^2$. We point out that the reason of the effect on Θ_1 of the observable variables of the second process $\{Y_{2t}\}$, whose conditional densities only depend on Θ_2, is the existing correlation between the first level parameters Θ_1 and Θ_2.

As concerns the second level parameter Λ, its posterior density has the same form as in the preceding number:

$$\Lambda / \mathbf{y}_1, \mathbf{y}_2 \sim N \left\{ \mu_0 + (\sigma_0^2 . \mathbf{1'}_{m+n}) . D^{-1} . \begin{bmatrix} y_1 - \mu . 1_m \\ y_2 - \mu . 1_n \end{bmatrix} \right. ;$$

$$\sigma_0^2 - (\sigma_0^2 . \mathbf{1'}_{m+n}) . D^{-1} . (\sigma_0^2 . \mathbf{1'}_{m+n})' \left. \right\} .$$

We still have $\text{Cov}(\Lambda, Y_{1t}) = \text{Cov}(\Lambda, Y_{2t}) = \sigma_0^2$ and then we could think that both observable processes exert the same influence on Λ. It is not so because of the difference in the second moments: $\text{Var}(Y_{it}) = \sigma_i^2 + s_i^2 + \sigma_0^2$ and $\text{Cov}(Y_{is}, Y_{it}) = s_i^2 + \sigma_0^2$, $i = 1, 2$.

All preceding considerations may be repeated, without important difference, for partially exchangeable models constituted by more than two subprocesses.

A final remark: in this expository paper, we have tried to illustrate that the usual hierarchical models lead to schemes of partial exchangeability for sequences of observable variables and that such schemes might be useful in simple problems involving longitudinal data. Obviously, we do not want to maintain that partially exchangeable models constitute the main tool in the realm of statistical meta-analysis; at most, they might be considered useful approximations for more complicated situations of interest.

References

Aldous, D. J. (1981) Representations for partially exchangeable random variables. *Journal of Multivariate Analysis,* 11, 581—598.

Bernardo, J. M. and Smith, A. F. M. (1995) *Bayesian Theory.* New York: J. Wiley & Sons.

Daboni, L. and Wedlin, A. (1982) *Statistica: un'introduzione alla impostazione neo-Bayesiana.* Torino: UTET.

Deely, J. J. and Lindley D. V. (1981) Bayes empirical Bayes. *J.A.S.A.,* vol.76, 833—841.

De Finetti, B. (1928) *Atti del Congresso Internazionale dei Matematici,* Bologna: Zanichelli.

De Finetti, B. (1937) Sur la condition d'equivalence partielle. *Colloque consacré a la theorié des probabilités.* Université de Genève, 12-16 octobre 1937.

De Finetti, B. (1938) Resoconto critico del Colloquio di Ginevra intorno alla Teoria delle probabilità. *Giornale dell' Instituto Italiano degli Attuari Roma.*

De Finetti, B. (1970) Teoria delle probabilità. Torino: Einaudi.

Diaconis, P. and Freedman, D. (1984) Partial exchangeability and sufficiency. In J. K. Ghosh and J. Roy (Eds.) *Statistics: Applications and New Directions.* Indian Statistical Institute, 205—236.

Good, I. J. (1965) *The Estimation of Probabilities. An Essay on Modern Bayesian Methods.* Cambridge, MA: MIT Press.

Lindley, D. V. (1971) The estimation of many parameters. In V. P. Godambe and D. A. Sprott (Eds.) *Foundations of Statistical Inference.* Toronto: Holt, Rinehart & Winston, 435—453.

Lindley, D. V. and Smith, A. F. M. (1972) Bayes estimates for the linear model, *J.R.S.S.,* B 34, 160—175.

Robert, C. P. (1994) *The Bayesian Choice.* New York: Springer-Verlag.

Schervish, M. J. (1995) *Theory of Statistics.* New York: Springer-Verlag.

Suppes, P. (1974) The measurement of belief. *J.R.S.S.,* B 36.

Walley, P. (1991) *Statistical Reasoning with Imprecise Probabilities.* Chapman & Hall.

12

Weak and Strong Reversibility of Causal Processes*

PATRICK SUPPES

Stanford University

1 Introduction

It is widely known, and often commented upon, that the basic equations of classical physics remain valid under a transformation from time t to $-t$. It is often also remarked that this invariance under a change of direction of time is completely contrary to ordinary experience. So the invariance of well-established classical physics, as well as other parts of physics, for example, relativistic mechanics and the Schrödinger equation in quantum mechanics, creates a natural philosophical tension about the nature of causal processes. The purpose of the present article is not to resolve in any complete way this tension between physics and ordinary experience, but to at least reduce the tension by introducing two concepts of reversibility. The first is that of *weak reversibility,* which is what is exemplified in the invariance of the equations of classical physics under time reversal. This is the *weak sense* of reversibility, because we can usually distinguish by observation whether a system of particle mechanics is running one way or the other. The appearance is not the same when we go to a reversal of time. Elementary examples would be measuring the velocity of a particle accelerating from rest, moving in a straight

---

*I am indebted to Ubaldo Garibaldi for several useful comments on an earlier draft.

Stochastic Causality.
Maria Carla Galavotti, Patrick Suppes and Domenico Costantini (eds.).

line and hitting an impenetrable wall. The picture of change of velocity, especially, will be very different under time reversal. Under such reversal, the particle has a very large velocity immediately, as motion begins, and will continue to decelerate until it finally reaches the state of rest, quite contrary to what would be observed in the usual direction of time. Of course, this difference is well accepted in discussions of the invariance of classical physics under time reversal.

The concept of *strong reversibility* introduced here, and used in various analyses of a variety of natural phenomena, including stochastic processes, has a much stronger condition for reversibility. A process, deterministic or stochastic, is said to be strongly reversible if we are unable to distinguish whether the process is running forward or backward. To put it in vivid terms often used, the backward movie, that is, running the film in reverse, of the observation of the process, is indistinguishable by observation from the forward movie. In what follows, these ideas are made more precise, especially for stochastic processes. I should mention that the concept of strong reversibility used here corresponds to what is often defined in the stochastic-process literature as simply *reversible*. So the condition given below for a Markov chain to be strongly reversible is the same as the standard concept to be found, for example, in Feller (1950, p. 342) for Markov chains.

Although many of the formal distinctions introduced in what follows are familiar in the literature of probability theory, especially of stochastic processes, it is my impression that the distinction introduced here has not been sufficiently emphasized, as a distinction, in the philosophical discussions of time reversibility. Its introduction is meant to reduce the tension I mentioned at the beginning, in the sense that there is a close alignment between ordinary experience not being strongly reversible, and, similarly, for many trajectories of idealized particles or other processes of importance in physics.

2   Weak Reversibility

As already remarked, it is well known that classical particle mechanics and relativistic particle mechanics have the property of being weakly reversible, i.e., the transformation changing the direction of the time always carries systems of classical particle mechanics into systems of classical particle mechanics and, correspondingly, for the relativistic case. Detailed proofs of these results are to be found in McKinsey and Suppes (1953) for classical particle mechanics and in Rubin and Suppes (1954) for relativistic particle mechanics. Because of their technical character they are summarized in the Appendix.

The proof that first-order Markov chains are carried into first-order Markov chains under time reversal is straightforward. Here is the elementary proof for first-order chains with a finite number of states. In the proof it is assumed that all the probabilities that occur in the denominators of expressions have probability greater than zero.

THEOREM 1. *All first-order Markov chains are weakly reversible.*

Proof (for finite-state, discrete-time processes):

$$
\begin{aligned}
P(i_{n-1} \mid j_n k_{n+1}) &= \frac{P(i_{n-1} j_n k_{n+1})}{P(k_{n+1} \mid j_n) P(j_n)} \\
&= \frac{P(k_{n+1} \mid j_n) P(j_n \mid i_{n-1}) P(i_{n-1})}{P(k_{n+1} \mid j_n) P(j_n)} \\
&= \frac{P(j_n i_{n-1})}{P(j_n)} \\
&= P(i_{n-1} \mid j_n).
\end{aligned}
$$

The situation in quantum mechanics is somewhat more complicated concerning weak reversibility. Certainly, the Schrödinger equation is so reversible. It is weakly reversible under transformations changing the direction of time. On the other hand, in the standard accounts, this is not true of the measurement process (von Neumann, 1955, Ch. 5).

Finally, from a common-sense standpoint, ordinary experience is certainly not weakly reversible. Time has a fixed direction and the empirical evidence in terms of human experience of this direction is overwhelming.

Forward ≠ Backward

People almost never walk up stairs backward. No races are run backward, etc. This is not to claim the "backward movies" violate laws of mechanics, only inductive laws and facts of experience, most especially the universally accepted nature of human memory:

unknown future ≠ known past.

Some definitions. Before going further, it is desirable to introduce some standard stochastic concepts, implicitly assumed in Theorem 1. First, let $\mathbf{X}(t)$ be a stochastic process such that for each time t, $\mathbf{X}(t)$ takes values in a finite set, a restriction imposed only for simplicity of exposition. Any finite family $\mathbf{X}(t_1), \mathbf{X}(t_2), \ldots, \mathbf{X}(t_n)$ has a joint probability distribution. Using an obvious simplification of notation, a process is (*first-order*) *Markov* if and only if for any times $t_1 < t_2, \ldots, t_n$,

(1) $$P(x_n \mid x_{n-1} \ldots, x_1) = P(x_n \mid x_{n-1}),$$

where the simplified notation is defined as follows:

$$P(x_n \mid x_{n-1}) = P(\mathbf{X}(t_n) = j_n \mid \mathbf{X}(t_{n-1}) = i_{n-1}).$$

A Markov process is *homogeneous* if the transition probabilities such as (1) do not depend on time, and it is *irreducible* if any state j can be reached from any other state in a finite number of steps. For a discrete-time homogeneous Markov process, I write the transition probabilities in several different, but useful notations

$$
\begin{aligned}
P_{ij} = p(i,j) &= P(j_n \mid i_{n-1}) = P(\mathbf{X}_n = j \mid \mathbf{X}_{n-1} = i) \\
&= P(\mathbf{X}(t_n) = j_n \mid \mathbf{X}(t_{n-1}) = i_{n-1}),
\end{aligned}
$$

all of these often used in the probability literature.

Let $P_{ii}(n)$ be the probability of returning to state i in n transitions. The *period* $r(i)$ of a state i is the greatest common divisor of the set of integers n for which $P_{ii}(n) > 0$. A state is *aperiodic* if it has period 1. Moreover, it is easy to show that if a Markov process is irreducible every state has the same period. So an aperiodic process is one in which every state has period 1.

Hereafter, when I refer to a Markov *chain*, I am assuming a Markov process that is finite-state, discrete-time, homogeneous, irreducible and aperiodic.

The definitions of being Markov, homogeneous and irreducible all apply without change to continuous-time processes, but there are some further conditions to impose to avoid physically unlikely cases. First, we require such a process to remain in any state for a positive length of time, and, second, the process cannot pass through an infinite sequence of states in a finite time. Corresponding to the transition probability for discrete-time processes, the *transition rate* for continuous-time processes is defined as:

$$
q_{ij} = \lim_{\tau \to \infty} \frac{P(X(t + \tau) = j \mid X(t) = i)}{\tau}, \quad i \neq j,
$$

and for definitional purposes, we set $q_{ii} = 0$. When I refer to a Markov *process,* I will mean a continuous-time one. Such a process remains in each state for a length of time exponentially distributed with the parameter $q(i)$ defined as:

$$
q(i) = \sum_j q_{ij},
$$

and when it leaves state i it moves to state j with the probability

$$
p_{ij} = \frac{q_{ij}}{q_i}.
$$

Note that in general

$$
\sum_j q_{ij} \neq 1,
$$

i.e., transition rates from i to another state need not add up to 1, as must be the case for transition probabilities from state i in the discrete-time processes. (A good discussion of such continuous-time processes may be found in Kelly (1979).)

3 Strong Reversibility

The intuitive idea of strong reversibility is easily characterized in terms of movies. A movie is strongly reversible if an observer cannot tell whether the movie is being run forward or backward. There is no perceptual difference in the two directions. Now, as our ordinary movie experience tells us at once, this is a pretty exceptional situation. On the other hand, there are important physical processes, a few of which we will discuss, that do have strong reversibility. An interesting case to begin with is this. When is a Markov chain strongly reversible? The right setting for this is to restrict ourselves at once to stationary, ergodic Markov chains. A Markov chain is *stationary* if its mean distribution is the same for all times. It is *ergodic* if it has a unique asymptotic distribution independent of the initial distribution. Note that under these definitions, a Markov chain can be ergodic but not stationary if its initial distribution is different from its unique, asymptotic one. But here I assume ergodic chains are stationary.

An ergodic Markov chain is *strongly reversible* if

(2) $$\pi_i p_{ij} = \pi_j p_{ji}$$

for all states i and j, where π is the unique asymptotic distribution. The equations (2) are often called the *detailed balance* conditions for strong reversibility. Bernoulli processes, such as coin flipping, are strongly reversible processes, because there is no memory of past states: $p_{ij} = p_j$ & $p_{ji} = p_i$, so we satisfy (2). Moreover, any 2-state ergodic Markov chain is strongly reversible. Here is the proof. Since the process is ergodic

$$\begin{aligned} \pi_1 &= \pi_1 p_{11} + \pi_2 p_{21} \\ &= \pi_1(1 - p_{12}) + \pi_2 p_{21}, \end{aligned}$$

so

$$\pi_1 p_{12} = \pi_2 p_{21}.$$

On the other hand, many 3-state chains are not strongly reversible. Here is an example of a three-color spinner.

	B	R	Y
B	$\frac{1}{2}$	$\frac{1}{4}$	$\frac{1}{4}$
R	$\frac{3}{8}$	$\frac{1}{2}$	$\frac{1}{8}$
Y	$\frac{1}{8}$	$\frac{3}{8}$	$\frac{1}{2}$

We have at once for the mean distributions (π_b, π_r, π_y) of the three states:

$$\begin{aligned} \pi_b &= \tfrac{1}{2}\pi_b + \tfrac{3}{8}\pi_r + \tfrac{1}{8}(1 - \pi_b - \pi_r) \\ \pi_r &= \tfrac{1}{4}\pi_b + \tfrac{1}{2}\pi_r + \tfrac{3}{8}(1 - \pi_b - \pi_r); \end{aligned}$$

solving the two equations, we find:

$$\pi_b = \frac{13}{37}, \quad \pi_r = \frac{14}{37}, \quad \pi_y = \frac{10}{37}.$$

Here is the proof that it is not strongly reversible:

$$\pi_b p_{br} = \frac{13}{17} \cdot \frac{1}{4} \neq \frac{14}{37} \cdot \frac{3}{8} = \pi_r p_{rb}.$$

We can summarize what we have been saying about Markov chains in the following theorem, where in the notation of the theorem, $\overset{d}{=}$ means equal in distribution.

THEOREM 2. *If an ergodic Markov chain is strongly reversible, then it is impossible to distinguish whether a movie of it is running forward or backward, that is, any finite sequence of random variables of the process has the same distribution when the order of the random variables is reversed:*

$$(\mathbf{X}_0, \mathbf{X}_1, \ldots \mathbf{X}_n,) \overset{d}{=} (\mathbf{X}_n, \mathbf{X}_{n-1}, \ldots \mathbf{X}_0,)$$

I give the proof for $n = 2$:

$$
\begin{aligned}
P(\mathbf{X}_0 = i, \mathbf{X}_1 = j) &= P(\mathbf{X}_1 = j | \mathbf{X}_0 = i) P(\mathbf{X}_0 = i) \\
&= \pi_i p_{ij} \\
&= \pi_j p_{ji} \quad \text{by (2)} \\
&= P(\mathbf{X}_1 = i | \mathbf{X}_0 = j) P(\mathbf{X}_0 = j) \\
&= P(\mathbf{X}_1 = i, \mathbf{X}_0 = j).
\end{aligned}
$$

The concepts introduced hold for finite-state continuous-time changes by replacing transition probabilities by transition rates $q_{ij}, i \neq j$ with $q_{ij} \geq 0$, as defined above. The detailed balance conditions defining strong reversibility are the same in form as those of (2) with the $\pi_i \geq 0$ and summing to 1, as before:

$$(3) \qquad\qquad \pi_i q_{ij} = \pi_j q_{ji}.$$

With a proof just like that of Theorem 2, we have at once:

THEOREM 3. *If an ergodic continuous-time Markov process is strongly reversible, then it is impossible to distinguish whether a movie of it is running forward or backward.*

In the present context, there is nothing special about first-order Markov chains. A second-order ergodic Markov chain is strongly reversible iff

$$(4) \qquad\qquad \pi_i \pi_{ij} p_{ijk} = \pi_k \pi_{kj} p_{kji},$$

where a chain is second-order Markov if

$$P(x_n \mid x_{n-1}, \ldots, x_1) = P(x_n \mid x_{n-1}, x_{n-2}),$$

corresponding to (1) for first-order.

THEOREM 4. *If a second-order Markov chain is ergodic and strongly reversible, then it is impossible to distinguish a movie of it running forward or backward, i.e.,*

$$(\mathbf{X}_0, \mathbf{X}_1, \ldots \mathbf{X}_n,) \stackrel{d}{=} (\mathbf{X}_n, \mathbf{X}_{n-1}, \ldots \mathbf{X}_0,).$$

Proof for $n = 3$:

$$P(\mathbf{X}_0 = i, \mathbf{X}_1 = j, \mathbf{X}_2 = k)$$
$$= P(\mathbf{X}_2 = k | \mathbf{X}_1 = j, \mathbf{X}_0 = i)P(\mathbf{X}_1 = j | \mathbf{X}_0 = i)P(\mathbf{X}_0 = i)$$
$$= p_{ijk}\pi_{ij}\pi_i$$
$$= \pi_k\pi_{kj}p_{kji} \quad \text{by (4)}$$
$$= P(\mathbf{X}_2 = i | \mathbf{X}_1 = j, \mathbf{X}_0 = k)P(\mathbf{X}_1 = j | \mathbf{X}_0 = k)P(\mathbf{X}_0 = k)$$
$$= P(\mathbf{X}_2 = i, \mathbf{X}_1 = j, \mathbf{X}_0 = k).$$

As is obvious, this proof also works for continuous-time second-order Markov processes. Without too much effort, this same sort of proof can be extended to chains of infinite order that are ergodic and stationary. The line of reasoning required is developed in Lamperti and Suppes (1959).

Turning again from Markov chains to continuous-time Markov processes, simple examples of strongly reversible processes are ergodic *birth and death* processes, defined by the transition rates being zero except for $q(i, i+1) > 0$, representing a birth, and $q(i, i - 1) > 0$ representing a death. The detailed balance conditions (3) then assume the form

$$(5) \qquad\qquad \pi_i q_{i,i+1} = \pi_{i+1} q_{i+1,i},$$

where the π_i's form the stationary probability distribution of the finite set of states.

Ehrenfest model. A simplified model of statistical mechanics nicely exemplifies a birth and death process that is strongly reversible. In the spirit of Maxwell's demon, there are two containers of particles, thought of as ideal gas molecules. Let $\mathbf{X}(t)$ be the number of particles in container I, and, so, $k - \mathbf{X}(t)$ is the number of particles in container II. The "birth and death" process corresponds to increasing or decreasing the number of particles in container I, with the transition rates being:

$$q_{i,i-1} = i\lambda, \qquad\qquad i = 1, 2, \ldots, k,$$
$$q_{i,i+1} = (k - i)\lambda, \qquad i = 0, 1, \ldots, k - 1,$$

where λ is the rate parameter. The stationary distribution may be shown to be

$$\pi_i = 2^{-k}\binom{k}{i} \qquad i = 0, 1, \ldots, k,$$

where
$$\binom{k}{i} = \frac{k!}{(k-i)!i!} \text{ and } 0! = 1.$$

That the process is strongly reversible may be seen by checking the detailed balance conditions (5).

$$
\begin{aligned}
\pi_i q_{i,i+1} &= 2^{-k}\binom{k}{i}(k-i)\lambda \\
&= \frac{2^{-k}k!(k-i)\lambda}{(k-i)!i!} \\
&= \frac{2^{-k}k!\lambda}{(k-(i+1))!i!} \qquad since \frac{k-i}{(k-i)!} = \frac{1}{(k-(i+1))!} \\
&= \frac{2^{-k}k!(i+1)\lambda}{(k-(i+1))!(i+1)!} \qquad since \frac{1}{i!} = \frac{i+1}{(i+1)!} \\
&= 2^{-k}\binom{k}{i+1}(i+1)\lambda \\
&= \pi_{i+1}q_{i+1,i}.
\end{aligned}
$$

Without attempting a detailed discussion of entropy for this process, I note that even though it is stationary and strongly reversible, the mean fluctuation from moment to moment is always stronger toward the equal distribution of particles in the two containers rather than away from this equal distribution, for

$$q_{i,i-1} = i\lambda < (k-i)\lambda = q_{i,i+1}$$

if and only if $i > \frac{k}{2}$, independent of the rate parameter λ. If we compute the entropy by the relative frequency $f(t)$ at time t of the number of particles in container I, then the "momentary" entropy of the process at t is

$$H(t) = -(f(t)\log f(t) + (1-f(t))\log(1-f(t))$$

and by the analysis just given the mean rate of change of entropy is always increasing, i.e., is positive, in spite of the strong reversibility of the process. But there is, obviously, no contradiction between these two features of the model, even at equilibrium, i.e., when stationary. An excellent detailed analysis of the Ehrenfest model and also of the Boltzmann equation, and of the associated "paradoxes" of Loschmidt and Zermelo, is to be found in Kac (1959).

Deterministic systems. I now turn to deterministic classical physical systems. It should be evident enough that only very special ones are strongly reversible. Any sort of dissipative system will, in general, not be. A real billiard ball on a real table is not strongly reversible. We start the motion and it comes to a halt somewhere, due to friction on the table. We can easily distinguish the forward from the backward picture. On the other hand, an idealized billiard ball on an idealized table with total conservation of energy and exact satisfaction of the equality of the angle of incidence and the angle of reflection can be put in periodic motion and, once in motion, will continue forever. Moreover, this idealized motion will, of course, be strongly reversible. By "observing" it, we cannot decide whether we are seeing the forward or the backward movie. (I put "observing" in quotes, for there is, in fact, no such billiard case, but we can artificially simulate it well for a finite period of time.)

A simple, but important, example of a classical system that has strong reversibility is an undamped and undriven one-dimensional harmonic oscillator. The differential equation for such an oscillator is:

$$\frac{d^2x}{dt^2} + \omega^2 x = 0,$$

where ω is the natural frequency of the oscillator when undamped and undriven, and the "initial" conditions at time $t = 0$ are:

$$x_0 = \alpha,$$
$$\frac{dx_0}{dt} = 0.$$

(When the model is a pendulum, the initial conditions correspond to the pendulum being at rest at time $t = 0$ with displacement α.) The general solution of (1) is:

$$x(t) = A\cos\omega t + B\sin\omega t,$$

and

$$\frac{dx}{dt} = -A\omega\sin\omega t + B\omega\cos\omega t$$

so at $t = 0$, $B = 0$, whence

$$A\cos\omega t = \alpha \text{ at } t = 0, \text{ and } A = \alpha.$$

Then

$$\frac{d^2x}{dt} = -A\omega^2\cos\omega t$$

and finally

(6) $$x(t) = A\cos\omega t.$$

We have at once that such an oscillator is strongly reversible, since for all t

$$x(-t) = A\cos-\omega t = A\cos\omega t = x(t).$$

Notice that this solution holds for $-\infty < t < \infty$, and the conditions at $t = 0$ are not really initial conditions, but just conditions for some t that yield a solution simple in form.

On the other hand, the damped, undriven harmonic oscillator, whose differential equation is:

$$\frac{d^2 x}{dt^2} + 2k\frac{dx}{dt} + \omega^2 x = 0,$$

where k is the damping coefficient, is easily shown not to be strongly reversible. This result is scarcely surprising, since a damped oscillator (in one dimension) is one of the simplest physical examples of a dissipative system. In the standard case where the damping is not too heavy, the oscillating function decreases in amplitude with time. More specifically the envelope of the oscillations is a negative exponential of the form e^{-kt} on the positive side and $-e^{-kt}$ on the negative side.

Without question, if the aim is to find examples of systems in classical physics that are not strongly reversible, and, therefore, the causal analysis is not strongly reversible, the place to look is among the many kinds of dissipative systems. The great prevalence of such systems in nature reinforces our personal experience to support the common-sense view that, as the saying goes, "... of course time is not reversible. Who could ever think otherwise?"

APPENDIX

A Classical Mechanics

A.1 Primitive Notions

The axiomatization of particle mechanics is based on five primitive notions: P, T, m, s, and f. P and T are sets, m is a unary function, s is a binary function, and f is a ternary function. The intended physical interpretation of P is as the set of particles. T is to be interpreted physically as a set of real numbers measuring elapsed times (in terms of some unit of time, and measured from some origin of time). If p is a member of P (that is to say, in the physical interpretation, if p is a particle), then $m(p)$ is to be interpreted physically as the numerical value of the *mass* of p. If p is in P, and t is in T, then $s(p, t)$ is an n-dimensional vector. For $n = 3$ (or for $n < 3$, if we are concerned with plane particle mechanics, or with one-dimensional particle mechanics) $s(p, t)$ is to be interpreted physically as a vector giving the *position* of p at time t. The primitive s fixes the choice of a coordinate system.

A.2 Axioms

Definition A1. *A structure* $\mathcal{P} = (P, T, m, s, f)$ *is an* (*n-dimensional*) *system of classical particle mechanics if and only if it satisfies the following axioms:*

Kinematical Axioms

Axiom P1. *P is a non-empty, finite set.*

Axiom P2. *T is an interval of real numbers.*

Axiom P3. *If p is in P and t is in T, then s(p,t) is an n-dimensional vector such that $\frac{d^2}{dt^2}s(p,t)$ exists.*

Dynamical Axioms

Axiom P4. *If p is in P, then m(p) is a positive real number.*

Axiom P5. *If p is in P and t is in T, then f(p,t,1),f(p,t,2), ...f(p,t,i), ... are n-dimensional vectors such that the series $\sum_{i=1}^{\infty} f(p,t,i)$ is absolutely convergent.*

Axiom P6. *If p is in P and t is in T, then*

$$m(p)\frac{d^2}{dt^2}s(p,t) = \sum_{i=1}^{\infty} f(p,t,i).$$

The omission of Newton's third law is deliberate in this general formulation. I specialize the axioms further below. But the level of generality of Definition A1 is useful for two purposes. First, a more general invariance theorem than is standard can be proved, as can be seen from the statement of the next theorem. Secondly, this general form of the axioms is most directly comparable to the axioms given below for relativistic particle mechanics, for which analogues of Newton's third law do not exist, because of the noninvariance of simultaneous distant events in the relativistic setting.

THEOREM A1. *Let $\mathcal{P} = (P,T,m,s,f)$ be an n-dimensional system of classical particle mechanics; let $\mathcal{A}$ be a nonsingular square matrix of order n; let $\mathcal{B}$ and $\mathcal{C}$ be n-dimensional vectors; and let α, β, and γ be real numbers such that $\beta \neq 0$ and $\gamma > 0$. Let T' be the set of all real numbers t' such that, for some t in T,*

(i) $$t' = \alpha + \beta t,$$

and let m', s', and f' be defined by the following equations, for p any element of P,t', any element of T', and i any member of I:

$$m'(p) = \gamma m(p)$$

(ii) $$s'(p,t') = \beta^2 s\left(p, \tfrac{t'-\alpha}{\beta}\right) \cdot \mathcal{A} + t'\mathcal{B} + \mathcal{C}$$

$$f'(p,t',i) = \gamma f\left(p, \frac{t'-\alpha}{\beta}, i\right) \cdot \mathcal{A}.$$

Then $\mathcal{P} = (P,T',m',s',f')$ is also an n-dimensional system of classical particle mechanics. Conversely, if equations (i) and (ii) hold, and (P,T',m',s',f') is an n-dimensional system of particle mechanics, then so is (P,T,m,s,f).

The surprise in Theorem A1 is that the matrix A is not restricted to a uniform change in the unit of measurement of spatial distance by necessarily being a similarity matrix, but is only nonsingular, which makes it a general affine matrix permitting different changes in spatial units along different dimensions.

Let ϕ_1 be a function which maps R (the set of all real numbers) onto itself in a one-to-one way; let ϕ_2 be a real-valued function defined over the set of positive real numbers; let ϕ_3 be a function which is defined over $E_n \times R$ (where E_n is the set of all n-dimensional vectors), and whose values are in E_n; and let ϕ_4 be a function which maps E_n onto itself in a one-to-one way. Then we call the ordered quadruple $(\phi_1, \phi_2, \phi_3, \phi_4)$ a *classically eligible transformation*.

DEFINITION A2. *Let* $\Phi = (\phi_1, \phi_2, \phi_3, \phi_4)$ *be a classically eligible transformation, and let* $\mathcal{P} = (P, T, m, s, f)$ *be a system of particle mechanics. Then the* Φ-*transform of* $\mathcal{P}$ *is the structure* (P, T', m', s', f'), *where* T' *is the set of all real numbers* t' *such that, for some* t *in* T,

$$t' = \phi_1(t);$$

and m', s', *and* f' *are defined by the following equations (for p any element of* P, t' *any element of* T', *and* i *any element of* I):

$$\begin{aligned} m'(p) &= \phi_2[m(p)] \\ s'(p, t') &= \phi_3[s(p, \phi_1^{-1}(t')), t'] \\ f'(p, t', i) &= \phi_4[f(p, \phi_1^{-1}(t'), i)]. \end{aligned}$$

An important, but not necessary, restriction on φ_1 of an eligible transformation is that it holds uniformly as the transform of the measure of time for all particles p in P. Since the general axioms of Definition A1 do not postulate any kind of interaction between the particles, a separate time measure could have been introduced for each particle, so that we would have an indexing of φ_1 on P:

$$t' = \varphi_{1,p}(t),$$

and then the measure of time could be reversed for an arbitrary subset of P. Our subsequent imposition of Newton's third law prevents this, but it raises a problem, discussed below, for relativistic mechanics, for which simultaneous action at a distance, as by gravity or classical electrostatic attraction or repulsion, is not invariant under the Lorentz transformations.

THEOREM A2. *Let* $\Phi = (\phi_1, \phi_2, \phi_3, \phi_4)$ *be an eligible transformation, and suppose that, for every n-dimensional system of classical particle mechanics* $\mathcal{P}$, *the* Φ-*transform of* $\mathcal{P}$ *is again a system of classical particle mechanics. Then there are real numbers* α, β, γ, *with* $\beta \neq 0$, *and* $\gamma > 0$, *n-dimensional vectors* $\mathcal{B}$ *and* $\mathcal{C}$, *and a nonsingular square matrix* $\mathcal{A}$, *of order n, such that, for x and Z any members of R and* E_n *respectively, and for y any positive member of R,*

$$\begin{aligned}
\phi_1(x) &= \alpha + \beta x \\
\phi_2(y) &= \gamma y \\
\phi_3(Z, x) &= \beta^2 Z \cdot \mathcal{A} + x\mathcal{B} + \mathcal{C} \\
\phi_4(Z) &= \gamma Z \cdot \mathcal{A}.
\end{aligned}$$

We now modify the axioms of Definition A1 to introduce Newton's third law—the new axioms are those given in Suppes (1957/1999). New Axiom P6 requires that the internal force $g(p, q, t)$ of particle q on particle p at time t is equal and opposite to $g(q, p, t)$. New Axiom P7 requires that the direction of interaction of the internal forces between particles p and q be the same as the direction of the vector $s(p, t) - s(q, t)$ from the position of p at time t to the position of q at time t. In stating P7 it is standard to use the vector or cross product x of two vectors (not to be confused with the Cartesian product of sets). Thus

$$\begin{aligned}
x \times y &= (x_1, x_2, x_3) \times (y_1, y_2, y_3) \\
&= (x_2 y_3 - x_3 y_2, x_3 y_1 - x_1 y_3, x_1 y_2 - x_2 y_1).
\end{aligned}$$

Note that the direction of the vector product is perpendicular to the plane formed by the vectors x and y. We also restrict the internal forces (Axiom P8) below to be functions only of the distance between particles. Finally, in addition to introducing the concept of internal force $g(p, q, t)$, we must modify Newton's second law, Axiom P6 of Definition A1, to include internal forces in new Axiom P9 below.

Definition A3. *A structure* $\mathcal{P} = (P, T, m, s, f, g)$ *is a system of ultraclassical particle mechanics if and only if it satisfies the following axioms:*

Axioms P1-P5. *Same as Definition A1.*
Axiom P6. *For p and q in P and t in T,*

$$g(p, q, t) = -g(q, p, t).$$

Axiom P7. *For p and q in P and t in T,*

$$s(p, t) \times g(p, q, t) = -s(q, t) \times g(q, p, t).$$

Axiom P8. *Any internal force $g(p, q, t)$ is a function only of the Euclidean distance between the particles p and q at time t.*
Axiom P9. *For p in P and t in T,*

$$m(p) \frac{d^2}{dt^2} s(p, t) = \sum_{q \in P} g(p, q, t) + \sum_{i=1}^{\infty} f(p, t, i).$$

The following two theorems are about such systems of ultraclassical particle mechanics. Together they show that once interacting forces satisfying

P6-P8 are assumed, then the standard transformation result, with $\mathcal{A}$ a similarity matrix, obtains.

THEOREM A3. *If the matrix $\mathcal{A}$ of Theorem A1 is a similarity matrix, then $\mathcal{P}'$ is ultraclassical if and only if $\mathcal{P}$ is ultraclassical.*

THEOREM A4. *Let $\Phi = (\phi_1, \phi_2, \phi_3, \phi_4)$ be an eligible transformation in the sense of Definition A2, and for ultraclassical particle mechanics, let φ_4 be extended to the internal forces: $g'(p, q, t') = \varphi_4[f(p, q, \varphi^{-1}(t')]$ and suppose, moreover, that the Φ-transform of every ultraclassical system is ultraclassical. Then the matrix $\mathcal{A}$ of Theorem A2 is a similarity matrix.*

B Relativistic Mechanics

The primitive concepts of relativistic mechanics are the same as those used in Definition A1, with the addition of the positive constant c, standing physically for the speed of light.

Definition B1. *A structure $\mathcal{P} = (P, T, m, s, f, c)$ is an (n-dimensional) system of relativistic particle mechanics if and only if it satisfies the following axioms:*

Kinematical Axioms

Axioms P1-P3. *Same as Definition A1.*

Axiom P4. *The constant c is a positive real number such that for every p in P and t in T*

$$|\frac{d}{dt} s_p(t)| < c.$$

Axioms P5 and P6. *Same as axioms P4 and P5 of Definition A1.*

Dynamical Axiom

Axiom P7. *If $p \in P$ and $t \in T$, then*

$$m(p) \frac{d}{dt} \left[\frac{\frac{ds_p(t)}{dt}}{(1 - \frac{|v_p(t)|^2}{c^2})^{\frac{1}{2}}} \right] = \left(1 - \frac{|v_p(t)|^2}{c^2} \right)^{\frac{1}{2}} \sum_{i=1}^{\infty} f(p, t, i).$$

As Axiom P7 makes clear, the concept of rest mass, $m(p)$, is taken as primitive, and the relativistic mass is then definable.

We next introduce the concept of a generalized Lorentz matrix, corresponding to the affine matrix $\mathcal{A}$ in the transformation theorems for classical mechanics. The term "generalized" means that changes in units of measurement can be made in transforming from one relativistic mechanical system to another, and thus both c and c', for the speed of light are needed. The positive number λ used with c and c' represents a uniform stretch, if $\lambda > 1$, or shrinking, if $\lambda < 1$, of space and time. As a later theorem shows this holds in general for eligible transformations of the relativistic systems, a more restricted result than that for classical systems in the sense of Definition A1. The n-dimensional vector U is to be interpreted as the velocity of the second frame of reference with respect to the first one, in the notation used below.

Let c, c', and λ be positive real numbers. Then a matrix $\mathcal{A}$ of order $n + 1$ is said to be a *generalized Lorentz matrix with respect to* (c, c', λ) if and only if there exist numbers δ and β, an n-dimensional vector U, and an orthogonal matrix $\mathcal{E}$ of order n, such that

$$\delta^2 = 1, \qquad \beta^2 \left(1 - \frac{U^2}{c'^2} \right) = 1,$$

and

(i) $$\mathcal{A} = \lambda \begin{pmatrix} \mathcal{I} & 0 \\ 0 & \frac{c}{c'} \end{pmatrix} \begin{pmatrix} \mathcal{E} & 0 \\ 0 & \delta \end{pmatrix} \begin{pmatrix} \mathcal{I} + \frac{\beta-1}{U^2} U^*U & -\frac{\beta U^*}{c'^2} \\ -\beta U & \beta \end{pmatrix}.$$

The following two lemmas simplify the statement and proof of Theorem B1.

Lemma B1. *Let* $(\{1\}, T, m, s, f, c)$ *be a system of relativistic particle mechanics, let c' and λ be positive real numbers, and let $\mathcal{A}$ be a generalized Lorentz matrix with respect to (c, c', λ). Let the function h be defined by the equation (for every t in T):*

$$h(t) = [(s_1(t), t)\mathcal{A}]_{n+1}.$$

Then the function $\frac{dh}{dt}$ exists; its values are either always positive or always negative; and the function h is one-to-one.

The following theorem of matrix theory provides a more elegant characterization of generalized Lorentz matrices.

Lemma B2. *Let c, c', and λ be positive real numbers. Then a matrix $\mathcal{A}$ of order $n + 1$ is a generalized Lorentz matrix with respect to (c, c', λ) if and only if*

(i) $$\mathcal{A} \begin{pmatrix} \mathcal{I} & 0 \\ 0 & -c'^2 \end{pmatrix} \mathcal{A}^* = \lambda^2 \begin{pmatrix} \mathcal{I} & 0 \\ 0 & -c^2 \end{pmatrix}.$$

THEOREM B1. *Let (P, T, m, s, f, c) be an n-dimensional system of relativistic particle mechanics. Let c', γ, and λ be positive real numbers, let B be an $(n + 1)$-dimensional vector, and let $\mathcal{A}$ be a generalized Lorentz matrix with respect to (c, c', λ). Let the function h be defined as follows (for all t in T):*

$$h(t) = [(s_p(t), t)\mathcal{A} + B]_{n+1}.$$

(By Lemma B1 the inverse function h^{-1} exists.) Let the function T' be defined as follows: T' is the range of the function h; and let the functions m', s', and f' be defined by the following equations (for p in P, t' in T and i in I):

$$m'(p) = \gamma m(p),$$
$$s'(p, t') = [(s(p, h^{-1}(t')), h^{-1}(t'))\mathcal{A} + B]_{1,\dots,n},$$
$$f'(p, t', i) = \frac{\gamma c'^2}{\lambda^2 c^2}\left[\left(f(p, h^{-1}(t'), i), \frac{f(p, h^{-1}(t'), i) \cdot v_p(h^{-1}(t'))}{c^2}\right)\mathcal{A}\right]_{1,\dots,n}.$$

Then $\mathcal{P}' = (P, T', m', s', f', c')$ is an n-dimensional system of relativistic particle mechanics.

Let ϕ_1 be a function mapping R^+ into R^+; let ϕ_2 be a function which is a one-to-one mapping of E_{n+1} into itself; and let ϕ_3 be a function mapping E_{2n} into E_n. Then we call the ordered triple (ϕ_1, ϕ_2, ϕ_3) a *relativistically eligible transformation*.

Definition B2. *Let $\Phi = (\phi_1, \phi_2, \phi_3)$ be a relativistically eligible transformation; let $\mathcal{P} = (P, T, m, s, f, c)$ be a system of relativistic particle mechanics; and let the function H be defined as follows (for every t in T):*

$$H(t) = [\phi_2(s(p, t), t)]_{n+1}.$$

Then by the Φ-transform of P (which we also write: $\Phi(P)$), we mean the structure (P, T', m', s', f'), where for p in P:

$$m'(p) = \phi_1(m(p));$$

T is the range of the function H; and s' and f' are defined by the following equations for t' in T', if the pre-image $H^{-1}(t')$ of t' under H is unique, and otherwise they are undefined:

$$s'(p, t') = [\phi_2(s(p, H^{-1}(t'))), H^{-1}(t')]_{1,\dots,n}$$

$$f'(p, t', i) = \phi_3(f(p, H^{-1}(t'), i)v(p, H^{-1}(t'))),$$

for $i \geq 1$.

THEOREM B2. *Let $\Phi = (\phi_1, \phi_2, \phi_3)$ be an eligible transformation, and let c and c' be positive real numbers such that (i) for every n-dimensional system of relativistic particle mechanics $\mathcal{P}_c$, $(\Phi(\mathcal{P}_c), c')$ is a system of relativistic particle mechanics, and (ii) ϕ_2 carries no c-line into a c'-particle path. Then there exist positive real numbers γ and λ, an $(n+1)$-dimensional vector B, and a generalized Lorentz matrix $\mathcal{A}$ with respect to (c, c', λ), such that, for* *any vectors Z_1 and Z_2 in E_n with $|Z_2| < c$, every x in R, and y in R^+,*

$$
\begin{aligned}
\phi_1(y) &= \gamma y, \\
\phi_2(Z_1, x) &= (Z_1, x)\mathcal{A} + B, \\
\phi_3(Z_1, Z_2) &= \tfrac{\gamma c'^2}{\lambda^2 c^2} \left[\left(Z_1, \tfrac{Z_1 \cdot Z_2}{c^2}\right)\mathcal{A}\right]_{1,\dots,n}.
\end{aligned}
$$

Condition (ii) of the theorem just requires that φ_2 can map no light line in the first frame of reference into a possible particle path in the second frame of reference. The lengthy proof of this theorem, which is given in Rubin and Suppes (1954), arises from the fact that no assumption about the continuity of φ_2 and φ_3 of an eligible transformation is made.

For the framework of the main part of this article, three theorems in the appendix state fundamental results on weak, but not strong, reversibility of mechanical systems. Theorem A2 says that every classical system of particle mechanics is weakly reversible, i.e., can be transformed into a system of classical mechanics with time reversed ($\beta < 0$ in the notation of the theorem). Theorem A4 says the same thing for ultraclassical systems. Finally, in the definition of a generalized Lorentz matrix, just before Lemma A1, $\delta = -1$ represents time reversal, which is not as evident in the formulation of Theorem B2, as it was in the classical case, but this is just a matter of the technical details of formulation. On the matter of time reversal considered here, there is no conceptual difference between classical and relativistic systems of particle mechanics. In both cases the time reversal itself is restricted to a linear transformation.

References

Feller, W. (1950) *Probability Theory and its Applications.* New York: John Wiley & Sons.

Kac, M. (1959) *Probability and Related Topics in Physical Sciences.* New York: Interscience Publishers.

Kelly, F. P. (1979) *Reversibility and Stochastic Networks.* New York: John Wiley & Sons.

Lamperti, J. and Suppes, P. (1959) Chains of infinite order and their application to learning theory. *Pacific Journal of Mathematics,* **9**, 739-143.

McKinsey, J. C. C., Sugar, A. C. and Suppes, P. (1953) Axiomatic foundations of classical particle mechanics. *Journal of Rational Mechanics and Analysis,* **2**, 253-272.

McKinsey, J. C. C. and Suppes, P. (1953) Transformations of systems of classical particle mechanics. *Journal of Rational Mechanics and Analysis,* **2**, 273-289.

Rubin, H. and Suppes, P. (1954) Transformations of systems of relativistic particle mechanics. *Pacific Journal of Mathematics,* **4**, 563-601.

Suppes, P. (1957) *Introduction to Logic*. Princeton: Van Nostrand Co. Reprinted 1999 by Dover Publications, New York.

Von Neumann, J. (1955) *Mathematical Foundations of Quantum Mechanics*. Princeton: Princeton University Press. Translated from the German edition of 1932 by Robert T. Beyer.

13

A Characterization of Creation and Destruction Propensities

DOMENICO COSTANTINI* AND UBALDO GARIBALDI[†]

*University of Bologna and [†] University of Geneva

1 Introduction

The need to assign probability values to single events is as old as statistical inference. As a matter of fact, the classical definition of probability concerns single events, and the first statistical inference, the Laplace rule of succession, was a way of assigning probability values to single events. This notwithstanding the problem of single events became crucial with quantum mechanics. "Some time before the discovery of quantum mechanics people realized that the connection between light waves and photons must be of a statistical character. What they did not clearly realize, however, was that the wave function gives information about the probability of *one* photon being in a particular place and not the probable number of photons in that place" (Dirac 1958, p. 9). Popper, in order to reform the objective statistical interpretation of probability in physics, suggests a new interpretation of probability, i.e., propensity: "what I propose is a *new physical hypothesis* (or perhaps a metaphysical hypothesis) analogous to the hypothesis of Newtonian forces. It is the hypothesis that every experimental arrangement [...] generates physical properties" (Popper 1959, p. 38). In his opinion "The idea of propensities is 'metaphysical' in exactly the same sense as forces or fields of forces are metaphysical. It is also 'metaphysical' in another sense: in the sense of pro-

Stochastic Causality.
Maria Carla Galavotti, Patrick Suppes and Domenico Costantini (eds.).
Copyright © 2001, CSLI Publications.

viding a coherent programme for physical research" (Popper 1957, p. 65). Recently Galavotti (1999) pointed out that this idea was not new: it had already been suggested by Campbell. Also for this author chance—this is the name he gave to probability—is a physical characteristic of the experimental arrangement.

It is not difficult to perceive not only in the idea but also in the very definition of propensity a metaphysical relish. Suppes pointed this out exactly "What troubles me [...] is the vagueness of his [of Popper's] new physical hypothesis in contrast to the sharpness of formulation of the hypothesis of Newtonian forces." (Suppes 1974, p. 768). In fact that proposed by Popper and other authors is no more than a new name for an old notion. This becomes evident if we recall that Popper maintained that the notion of probability can be interpreted in terms of the measure theory while he considered his interpretation to be very near the formal theory of probability. But a formal theory is compatible with a host of interpretations. This point has also been very clearly stressed by Suppes "In broad terms, the issue is that of characterizing the explicit meaning of the propensity interpretation" (Suppes 1974, p. 761). Fifteen years later, commenting with full lucidity on the R. N. Giere and D. H. Mellor contributions, he said "what is missing in these excellent intuitive discussions of the philosophical and scientific foundation of a propensity interpretation is any sense that there needs to be something proved about the propensity interpretation" (Suppes 1987, p. 335), and a few lines below "In order for an interpretation of probability to be interesting, some clear concepts need to be added beyond those in the formal theory as axiomatized by Kolmogorov. The mere hortatory remark that we can interpret propensity as probability directly [...] is to miss the point of giving a more thorough analysis" (Suppes 1987, p. 336).

It is precisely such an analysis that is lacking in Popper and his followers. In our opinion this is a crucial point. This lack can be fully grasped if one considers that all known probability interpretations give methods for determining probability values. This is not the case for the propensity interpretation. Nobody knows how a propensity can be determined. To say that propensity is a physical property of an experimental arrangement has no value in determining its values. In this sense, Popper is right in saying that the propensity interpretation is a program for physical research. As a matter of fact it is really no more than a program. The core of Suppes' criticism is that, apart from the contributions of Suppes himself, propensity remains no more than a research program.

The more thorough analyses for propensity that Suppes has performed have the form of representation theorems as he qualifies them. For example he has proved a representation theorem for: radioactive phenomena, the propensity to decay; psychological phenomena, the propensity of strength of response; gambling phenomena, the propensity of heads in coin tossing; random phenomena, the propensity of randomness. Perhaps the use of the term representation theorem is not really suitable. In our opinion what Suppes suggests is the characterization of probability distributions for the considered phenomena in terms of assumptions regarding: qualitative probability structures; choice probabilities; the mechanics of coin tossing considered as a classical mechanical system; the motion of the third particle with a nearly negligible mass in the three-body problem. Following the suggestion of Suppes, we propose a characterization, in Suppes' term a representation theorem, for creation and destruction propensities.

2   Destructions and Creations

Destructions and creations are the basic ingredient of any dynamics. In general terms, a system or a population is a collection of individuals belonging to different categories (bearing different properties). In the simplest case the dynamics of a system amount to a change in the number of individuals belonging to different categories. The most elementary dynamical event occurs when one individual changes category. Or, as we shall say, an individual belonging to one category leaves this category to go to another one. The most known example of such a change is the effect of an elastic collision of a particle against some fixed obstacle such as the Q-molecules of the famous example of the Ehrenfests or against the walls of a container. We shall call such an event an elementary transition, more specifically, such a transition happens when one individual leaves a category to go to another one. Of course, there are also binary, ternary and n-ary transitions. These are generalizations of the unary transitions, and what can be said about the most simple transitions can be repeated with minor changes for the more complex ones.

As we have seen transitions are considered in physics, but they also occur in other disciplines such as population genetics or economics. In the first case, to transitions the purely stochastic changes in gene frequency in finite populations have to be ascribed. For example, in population genetics a unary transition happens when a mutation occurs, that is when a gene changes its allele (Costantini and Garibaldi, forthcoming a); in economics, such a transition occurs when a shopper waiting at one shop, leaves this shop to go to another one (Costantini and Garibaldi, forthcoming b).

We shall consider a unary transition as the juxtaposition of two steps: a destruction and a creation in this order. A destruction in a category occurs when the number of individuals belonging to the categories decreases by one unit. Of course, after a destruction the number of individuals of the considered system decrease by one. But this decrement is purely virtual: the destroyed individual retires on a sort of pulpit and there he waits till he goes to another category. As we shall see, this virtual decrement is an essential step in the characterization of the transition probability. A creation in a category happens when the number of individuals belonging to the category increases by one unit. Thus the number of individuals in the system, which after a destruction decreased by one, as a consequence of a creation becomes that of the beginning again.

It follows that when we have both the destruction and the creation probabilities, we also have the probability of a unary transition. And when the transition probability gives rise to a homogeneous, ergodic and aperiodic Markov chain, it determines the equilibrium distribution seen as a probability distribution not changing with time. Briefly: by characterizing the destruction and the creation probabilities, we characterize the equilibrium distribution too.

As we shall see, the characterization of destruction probabilities simply amount to supposing the choice at random of one unit. The core of the problem is the characterization of creation probabilities. In reality such a characterization has already been given in geometrical terms by L. Brillouin (1927). What this author did can be seen as a way of representing the propensity of the cells to accommodate particles. According to Brillouin, at the outset cells have the same volume, but this changes according to their occupation numbers. Hence the accommodation propensity changes too. If a particle has a positive volume this amounts to negative correlation, a negative volume to positive correlation and no volume to statistical independence. But this cannot be done without taking a crucial peculiarity of Brillouin's geometrical model into account. In fact, when considering negative or null volumes it all works very well. On the contrary, in the case in which particles have positive volumes we are not sure that the probability axioms are always fulfilled. In fact, when all cells are already occupied, the addition rule does nor hold any longer. As a consequence we are compelled to consider new principles ensuring that creations and destructions are always ruled by probabilities. Essentially these principles are based on the concept of regularity to be defined in the next section.

3   Conditions for Regularity

Considering a system of n individuals and d categories, let $\mathrm{N}^{n,d}$ be the class of all d-tuples $\mathbf{n} := (n_1, ..., n_i, ..., n_d), n_i \geq 0, \sum_i n_i = n$. We say that $\mathbf{n}$ is the vector describing the state of the system, for short the vector of the system and n_i, the projection of $\mathbf{n}$ along the ith direction, is the occupation number of the ith category or the ith occupation number of $\mathbf{n}$. If $\mathbf{k}(i) := (0, ..., 1, ..., 0)$ is the i th versor, $\mathbf{n}^i := \mathbf{n} + \mathbf{k}(i) \in \mathrm{N}^{n+1,d}$ is the i-successor of $\mathbf{n}$ (the successor of $\mathbf{n}$ along the ith direction) and $\mathbf{n}_i := \mathbf{n} - \mathbf{k}(i) \in \mathrm{N}^{n-1,d}$ is the i-ancestor of $\mathbf{n}$ (the ancestor of $\mathbf{n}$ along the ith direction).

We say that $\mathbf{n}$ has a successor if for some i it has an i-successor. The same holds for ancestors. In general for each $\mathbf{n}$ there are successors and ancestors. But this is not always the case, for example $\mathbf{0}$, the sole member of $N^{0,d}$, has no ancestors. It is also worth considering a limiting case. This is the case in which the system has a maximum occupation vector or, what is the same, each cell has a maximum occupation number. We shall denote by $\mathbf{n}^*{:=}(n_1^*, ..., n_i^*, ..., n_d^*), n_i^* \geq 1, \sum_i n_i^* = n^*$, the maximum occupation vector and say that for the ith direction n_i^* is the maximum occupation number. The most important case is $\mathbf{1} :=(1, ..., 1, ..., 1)$, i.e., Pauli's occupation vector. Obviously n^* has no successors. It may happen that a d-tuple $\mathbf{n}$ only has $m < d$ successors; this is the case when for $d - m$ directions the occupation numbers are maximum

The system may change its size. When this is the case we say that the system undergoes an increment or a decrement. To describe such a transition we consider the vector $\mathbf{n}$ and its i-successor $\mathbf{n}^i$ or its ancestor $\mathbf{n}_i$. We denote by $\mathbf{n} \rightarrowtail \mathbf{n}^i$ the step from $\mathbf{n}$ to $\mathbf{n}^i$ and call it an i -increment or, in general, an increment. Likewise $\mathbf{n} \rightarrowtail \mathbf{n}_i$ is the step from $\mathbf{n}$ to $\mathbf{n}^i$ that we call an i -decrement or a decrement. An (increasing) pattern is a sequence of d-tuples such that the $(i + 1)$th term is a successor of the ith. Hence such a pattern is obtained by summing up a given $\mathbf{n}$ and a sequence of versors, $\mathbf{k}(i), \mathbf{k}(j), ..., \mathbf{k}(h)$. With the obvious changes, a decreasing pattern is defined in a similar way provided that no term of the sequence is $\mathbf{0}$. Now we are able to state precisely which steps are possible and which are not.

3.1 Regular Increments

Given an increment $\mathbf{n} \rightarrowtail \mathbf{n}^i$ we can define an increment function $R(\mathbf{n} \rightarrowtail \mathbf{n}^i) \in \{0,1\}$. The notion of regularity, which is the formal counterpart of that of a possible step, is defined in terms of this function. The increment $\mathbf{n} \rightarrowtail \mathbf{n}^i$ is regular if and only if $R(\mathbf{n} \rightarrowtail \mathbf{n}^i) = 1$. Using the notion of regular increment we can define that of regular pattern, that is we can say that a pattern is regular if and only if all its increments are regular.

We now take into account the conditions which define the class of regularity Re, that is the class of all regular increments. $\mathbf{n} \rightarrowtail \mathbf{n}^i \in$ Re is tantamount to $R(\mathbf{n} \rightarrowtail \mathbf{n}^i) = 1$.

R1) for all $i, \mathbf{0} \rightarrowtail \mathbf{0}^i \in$ Re,

R2) if $i \neq j$ and $\mathbf{n} \rightarrowtail \mathbf{n}^i, \mathbf{n} \rightarrowtail \mathbf{n}^j \in$ Re, then $\mathbf{n}^i \rightarrowtail \left(\mathbf{n}^i\right)^j \in$ Re,

where $\left(\mathbf{n}^i\right)^j := \mathbf{n}+ \mathbf{k}_i + \mathbf{k}_j$. These conditions are very simple, they only guarantee that: first, when the system is void all increments are regular; and, second, that the regularity of accommodations that do not take place is preserved. It can be shown (Costantini and Garibaldi, forthcoming b) that these conditions are enough to ensure the regularity of all patterns from $\mathbf{0}$ to $\mathbf{1}$.

In order to further characterize the propensity we need conditions fixing the values of the increment function when the "capacity" of the cells is taken into account. In this case what matters is the projection of $\mathbf{n}$ along the ith direction, i.e., n_i. To formalize this we suppose that

R3) if $\mathbf{n}'$ and $\mathbf{n}''$ are such that $n_i' = n_i''$, then $R(\mathbf{n}' \rightarrowtail \mathbf{n}'^i) = R(\mathbf{n}' \rightarrowtail \mathbf{n}''^i)$.

As a consequence with respect to regularity an i-increment can be written as $n_i \rightarrowtail n_i + 1$, that is $R(\mathbf{n} \rightarrowtail \mathbf{n}^i) = R(n_i \rightarrowtail n_i + 1)$.

We now consider the complementary class of Re, i.e. the class of all increments $\mathbf{n} \rightarrowtail \mathbf{n}^i$ such that $R(\mathbf{n} \rightarrowtail \mathbf{n}^i) = 0$. We shall denote this class with $\overline{\text{Re}}$, that is $\mathbf{n} \rightarrowtail \mathbf{n}^i \in \overline{\text{Re}}$ is tantamount to $R(\mathbf{n} \rightarrowtail \mathbf{n}^i) = 0$. At first we fix the limit of regularity, that is

R4) for all i, $n_i^* \rightarrowtail n_i^* + 1 \in \overline{\text{Re}}$.

Then we suppose that non regularity propagates according to the following condition

R5) for all i, if $\mathbf{n}$ is such that $n_i = n_i^*$, and $i \neq j$, then $\mathbf{n}^j \rightarrowtail (\mathbf{n}^i)^j \in \overline{\text{Re}}$.

The meaning of these two conditions is that $\mathbf{n}^*$ is the last d-tuple accessible to the system. As a consequence, all increments following a non regular one are non regular. This can be stated as:

THEOREM 1. *If* $\mathbf{n} \rightarrowtail \mathbf{n}^i \in \overline{\text{Re}}$, *then for all* j, $\mathbf{n}^j \rightarrowtail (\mathbf{n}^j)^i \in \overline{\text{Re}}$.

Of course, the maximum occupation number for the ith cell, n_i^*, is a free parameter. Moreover we are not compelled to make all the maximum occupation numbers equal, that is it is possible to fix $n_i^* \neq n_j^*$ for $i \neq j$. From the stated conditions it follows that all the patterns joining $\mathbf{0}$ to $\mathbf{n}^*$ are sole regular patterns. This is made apparent by the following theorem which shows a sort of protoexchangeability

THEOREM 2. *If* $R(\mathbf{n} \rightarrowtail \mathbf{n}^i) R(\mathbf{n}^i \rightarrowtail (\mathbf{n}^i)^j) > 0$, *then* $R(\mathbf{n} \rightarrowtail \mathbf{n}^j) R(\mathbf{n}^j \rightarrowtail (\mathbf{n}^j)^i) > 0$.

That is, if a two increment pattern is regular, then it can be reversed; in other words, if two increments are possible, they can happen in the reverse order.

The accommodation process may be pursued till the system is filled, that is till the system reaches $\mathbf{n}^*$, after that the process stops. From now on we only consider increments of the class Re. As a consequence all the patterns we are considering are regular, that is the final d-tuple is always accessible.

3.2 Regular Decrements

Given a vector $\mathbf{n}$, an i-decrement may happen if there is at least one individual belonging to the $i$the category.. This is the case if the ith occupation number of $\mathbf{n}$ is different from zero, i.e., $n_i > 0$. Hence the following is also a condition for regularity.

R6) if $n_i > 0$, then $\mathbf{n} \rightarrowtail \mathbf{n}_i \in \text{Re}$.

From this follows:

THEOREM 3. *If* $i \neq j, \mathbf{n} \rightarrowtail \mathbf{n}_i \in \text{Re}, \mathbf{n} \rightarrowtail \mathbf{n}_j \in \text{Re}$, *then* $\mathbf{n}_i \rightarrowtail (\mathbf{n}_i)_j \in$ Re, *where* $(\mathbf{n}_i)_j := \mathbf{n} - \mathbf{k}(i) - \mathbf{k}(j)$.

It is easy to show that all patterns joining $\mathbf{n}^*$ to $\mathbf{0}$ in both directions are regular. This individuates a class of regular (possible) patterns. The vectors in this class are joint by regular transitions. They are all the ancestors of $\mathbf{n}^*$. By writing $\mathbf{n} \in \text{Re}(\mathbf{n}^*)$ we denote a member of this class.

4   Conditions for Propensity

What we have been doing is a qualitative approach to transitions. A quantitative treatment takes place when we allot values, of the interval $[0, 1]$, to regular increments, i.e., when we take probabilities into account. In order to consider increments and decrements at the same time, we denote by $\mathbf{n}(i)$ both $\mathbf{n}^i$ and $\mathbf{n}_i$. We now consider the conditional probability $\Pr\{\mathbf{n}(i)|\mathbf{n}\}$. This is the probability that when the vector of the system is $\mathbf{n}$, one step will change the state into $\mathbf{n}(i)$. In order to make apparent the meaning of this conditional probability we shall write $\Pr\{\mathbf{n} \rightarrowtail \mathbf{n}(i)\}$. We prefer writing in this way because we want to emphasize the transformation determined by the step. The probability we are considering satisfies the following conditions:

P1) if $\mathbf{n}, \mathbf{n}(i) \in \text{Re}(\mathbf{n}^*)$, then $\Pr\{\mathbf{n} \rightarrowtail \mathbf{n}(i)\} > 0$;

P2) if $\mathbf{n}, \mathbf{n}(i) \in \text{Re}(\mathbf{n}^*)$, then $\sum_i \Pr\{\mathbf{n} \rightarrowtail \mathbf{n}(i)\} = 1$;

P3) if $\mathbf{n}, \mathbf{n}(i), \mathbf{n}(i,j) \in \text{Re}(\mathbf{n}^*)$, then $\Pr\{\mathbf{n} \rightarrowtail \mathbf{n}(i,j)\} = \Pr\{\mathbf{n} \rightarrowtail \mathbf{n}(i)\} \Pr\{\mathbf{n}(i) \rightarrowtail \mathbf{n}(i,j)\}$.

As we have said, all vectors considered in these conditions are regular. If this is the case, **P1**) ensures that increments and decrements have probability values and that these values are greater than zero. **P2** and **P3** are the addition and the product rules. A warning is necessary: in the d terms of **P2** all $\mathbf{n}(i)$ must be of the same type, that is all result either from d increments or from d decrements. On the other hand, no constraint is imposed on for the double step $\mathbf{n}(i, j)$ of **P3** that may regard two increments, two decrements or an increment and a decrement. A function is a regular probability if and only if it satisfies these three conditions.

4.1   Exchangeability

Some of the properties we are considering are consequences of the condition of exchangeability alone that we introduce as

P4) $\Pr\{\mathbf{n} \rightarrowtail \mathbf{n}(i, j)\} = \Pr\{\mathbf{n} \rightarrowtail \mathbf{n}(j, i)\}$,

where the two double steps are always supposed to be of the same sort, that is both increments or both decrements. This means that the probabilities of a double increment or of a double decrement do not change by reversing the order in which they happen. As we shall see, this is not the case for a decrement followed by an increment. A probability is regular and exchangeable if and only if it satisfies **P1, P2, P3** and **P4**.

For a regular, exchangeable probability we define the relevance quotient at **n**

$$Q_i^j(\mathbf{n}) := \frac{\Pr\{\mathbf{n}^j \rightarrowtail (\mathbf{n}^j)^i\}}{\Pr\{\mathbf{n} \rightarrowtail \mathbf{n}^i\}}$$

that, written as $\frac{\Pr\{\mathbf{n} \rightarrowtail (\mathbf{n}^i)^j\}}{\Pr\{\mathbf{n} \rightarrowtail \mathbf{n}^i\}\Pr\{\mathbf{n} \rightarrowtail \mathbf{n}^j\}}$, reveals itself as a way to measure (in)dependence.

4.2 Destruction Probability

The stated conditions ensure that all patters joining **n** to **0** are equiprobable.. This has a notable consequence that we want to stress by means of the following

THEOREM 4. *If* **P1, P2, P3** *and* **P4** *hold, then*

(1)
$$\Pr\{\mathbf{n} \rightarrowtail \mathbf{n}_i\} = \frac{n_i}{n}.$$

Proof. Immediate.

(1) is the probability of an i-decrement or, as one says in statistical mechanics, of destroying a particle of the ith cell, for simplicity of destroying in i, when there are n_i particles in it. We call (1) the destruction probability. Of course, starting from **n**, no more than n destructions can take place. Furthermore we note that if (1) holds, then all relevance quotients at **n** are equal. In fact

THEOREM 5. *If* **P1, P2, P3** *and* **P4** *hold, then all relevance quotients at* **n** *are equal.*

Proof. $\mathbf{Q}_i^j(\mathbf{n}) = \frac{\Pr\{\mathbf{n}_i \rightarrowtail (\mathbf{n}_i)_j\}}{\Pr\{\mathbf{n} \rightarrowtail \mathbf{n}_i\}} = \frac{n_j}{n-1}\frac{n}{n_j} = \frac{n}{n-1} = \frac{n_k}{n-1}\frac{n}{n_k} = \frac{\Pr\{\mathbf{n}_k \rightarrowtail (\mathbf{n}_k)_m\}}{\Pr\{\mathbf{n} \rightarrowtail \mathbf{n}_k\}} = \mathbf{Q}_m^k(\mathbf{n}).$

We call invariant a probability for which this property holds.

4.3 Creation Probability

In the application, especially in physics, there are two possibilities. In the first one, a maximum occupation vector exists for the system. In the second, such a vector does not exist, that is any number of individuals may belong to a category. At first we take into account the case in which there is a maximum occupation vector $\mathbf{n}^*$. The reason is that in this case it is possible to allot probability values to each increment using a regular, exchangeable probability which, obviously, depend upon $\mathbf{n}^*$. As is well known, the sole known physical case is that of the Pauli vector.

Maximum occupation vector

It is easy to realize that, when the system has a maximum occupation vector, then all patterns joining $\mathbf{0}$ to $\mathbf{n}^*$ have the same probability, and the unique representation of the increments probability is given by the following

THEOREM 6. *If* **P1, P2, P3** *and* **P4** *hold, then*

(2)
$$\Pr\{\mathbf{n} \rightarrowtail \mathbf{n}^i\} = \frac{n_i^* - n_i}{n^* - n}.$$

Proof. Immediate.

It is easy to check that if a probability function satisfies this theorem, then $\Pr\{\mathbf{n} \rightarrowtail \mathbf{n}^i\} \geq \Pr\{\mathbf{n}^i \rightarrowtail (\mathbf{n}^i)^i\}$. The equality sign regards the limiting case in which (2) is equal to 1. This case occurs when the individuals all belong to a sole category. In general, two transitions involving the same category are negatively correlated. The probability of an accommodation decreases with the number of particles in the cell and becomes zero when the cell is saturated.. Moreover, we have

THEOREM 7. *If* **P1, P2, P3** *and* **P4** *hold, then all relevance quotients at* n *are equal.*

Proof. $Q_i^j(\mathbf{n}) = \frac{\Pr\{\mathbf{n}^j \rightarrowtail (\mathbf{n}^j)^i\}}{\Pr\{\mathbf{n} \rightarrowtail \mathbf{n}^i\}} = \frac{n^* - n}{n^* - n - 1} = \frac{\Pr\{\mathbf{n}^k \rightarrowtail (\mathbf{n}^k)m\}}{\Pr\{\mathbf{n} \rightarrowtail \mathbf{n}^h\}} = Q_m^k(\mathbf{n})$.

This is an important consequence of exchangeability and of the existence of a maximum occupation vector. It is worth noting that, in the case of the existence of a maximum occupation vector, both the destruction probability and the creation probability are invariant. This property of the probability can in general be stated assuming that the relevance quotient is a function of n.

No maximum

Now we consider the case in which there is no maximum occupation d-tuple. If this is the case exchangeability is no longer enough for determining the probability of a transition. To this end we introduce the condition of invariance for regular, exchangeable probabilities.

P5) if $\mathbf{n}', \mathbf{n}'' \in \mathbf{N}^{n,d}$ and $i \neq j, h \neq k$, then $Q_{i}^{j}(\mathbf{n}) = Q_{h}^{k}(\mathbf{n})$.

We call invariant a probability for which **P5** holds. Invariance is a much stronger property than exchangeability. As a consequence of this condition we have

THEOREM 8. $\Pr\{\mathbf{n} \rightarrowtail \mathbf{n}^i\}$ *only depends upon* n_i *and* n.

Proof. For a fixed value of n we have $Q_i^j(\mathbf{n}) = Q_i^k(\mathbf{n})$ that is $\frac{\Pr\{\mathbf{n}^j \rightarrowtail (\mathbf{n}^j)^i\}}{\Pr\{\mathbf{n} \rightarrowtail \mathbf{n}^i\}}$ $= \frac{\Pr\{\mathbf{n}^k \rightarrowtail (\mathbf{n}^k)^i\}}{\Pr\{\mathbf{n} \rightarrowtail \mathbf{n}^i\}}$, from which $\Pr\{\mathbf{n}^j \rightarrowtail (\mathbf{n}^j)^i\} = \Pr\{\mathbf{n}^k \rightarrowtail (\mathbf{n}^k)^i\}$.

It is worth noting that this theorem is similar to **R3**). But, while regularity depends only on the occupation number, probability depends upon the occupation number, the total number of the considered particles and a parameter to be fixed. This is shown by:

THEOREM 9. *If* **P1, P2, P3, P4 and P5** *hold, then*

(3) $$\Pr\{\mathbf{n} \rightarrowtail \mathbf{n}^i\} = \frac{w_i + n_i}{w + n},$$

where $w_i := w p_i$, $p_i := \Pr\{\mathbf{0} \rightarrowtail \mathbf{0}^i\}$ *and* $w := \frac{Q_i^j(0)}{1 - Q_i^j(0)}$.

For the proof see Costantini and Garibaldi (1989). (2) is a special case of ((3) by putting $w_j = -n_j^*$ and $w = -n^*$ we have (2). When $p_i = d^{-1}$ for all i, (3) becomes $\frac{w d^{-1} + n_i}{w + n}$ that, by putting $c = dw^{-1}$, takes the form

(4) $$\Pr\{\mathbf{n} \rightarrowtail \mathbf{n}^i\} = \frac{1 + c n_i}{d + c n}$$

which was used by Brillouin. (4) represents all invariant accommodation processes. In fact, $c > 0$ represents positive correlation among particles; $c = 0$ represents independence, $c < 0, |c| = 1, 2, ...$, represents negative correlation among particles. In this case by putting $c = \frac{d}{n^*}$ we have (2).

4.4   Remarks on Destruction and Creation Probabilities

The difference between (1) and (4) is profound. The probability of destroying in a cell, depends only upon the number of particles that are in the cell, irrespective of the type of particles taken into account, that is it does not depend upon c. This probability is equal to the fraction of particles that are in the cell. In turn this holds if and only if all particles have the same probability of being destroyed irrespective of the cell in which they are. In other words, all particles are on a par with respect to destruction. Or, what is the same, the destruction of a particle always happens at random regardless of the cell in which the destruction takes place.

On the contrary, creation depends upon the interparticle correlation. That is, apart from the case in which $c = 0$, it depends upon the number of particles that are in the considered cell. The interparticle correlation is fixed by the value of c in (4). For $c = 1$, the particles are positively correlated and there is no limit to the number of particles that can be created in one cell. For $c = -1$, the particles are negatively correlated, and only one particle can be created in a cell.

For $c = 0$, the particles are stochastically independent. That is, when considering independent particles, the probability of creating a particle in a cell does not depend upon the number of particles already in the cell. This probability only depends upon the cell, and it is the reciprocal of the number of cells. In turn this holds if and only if all cells have the same probability of accommodating a particle irrespective of their occupation numbers. In other words, all cells are on a par with respect to creation, that is the choice of the cell is always at random.

Suppes, speaking of the characterization of an objective propensity, asserted that subjective theories of probability aim "to prove that the structural axioms impose a unique probability measure on events. It is this uniqueness that is missing from the objective theory [...], and in my own judgement this lack of uniqueness is a strength and not a weakness" (Suppes 1987, p. 337). This is exactly the case of (3). As we have seen, different values of the parameter c characterize the different behavior of elementary particles.

5 Transitions

When an individual belonging to the kth category leaves it and goes to the mth category, the state of the system undergoes a transition from $\mathbf{n} = (n_1, ..., n_k, ..., n_m, ..., n_d) \in \mathrm{N}^{n,d}$ to $\mathbf{n}_k^m := (n_1, ..., n_k - 1, ..., n_m + 1, ..., n_d) \in \mathrm{N}^{n,d}$. We denote this transition by $\mathbf{n} \Rightarrow \mathbf{n}_k^m$ considering it as the juxtaposition of two steps. That is $(\mathbf{n} \rightarrowtail \mathbf{n}_k)$ and $(\mathbf{n}_k \rightarrowtail \mathbf{n}_k^m)$, with $\mathbf{n}_k = (n_1, ..., n_k - 1, ..., n_m, ..., n_d) \in \mathrm{N}^{n-1,d}$. Thus, starting from $\mathbf{n}$ we can reach $\mathbf{n}_k^m$ passing through the intermediate (virtual) state $\mathbf{n}_k$. It is worth stressing that the two steps, though logically independent, are probabilistically dependent.

In order to determine the transition probabilities $\Pr\{\mathbf{n} \Rightarrow \mathbf{n}_k^m\}$, we appeal to (2) and (4). The probabilities of $\mathbf{n} \rightarrowtail \mathbf{n}_k$ and $\mathbf{n}_k \rightarrowtail \mathbf{n}_k^m$ are respectively

$$(5) \qquad \Pr\{\mathbf{n} \rightarrowtail \mathbf{n}_k\} = \frac{n_k}{n},$$

$$(6) \qquad \Pr\{\mathbf{n}_k \rightarrowtail \mathbf{n}_k^m\} = \begin{cases} \frac{w_m + n_m}{w + n} & \text{for } m \neq k \\ \frac{w_m + n_m - 1}{w + n} & \text{for } m = k \end{cases}.$$

Thanks to (5) and (6) and considering only the off-diagonal entries, for the probability of $\mathbf{n} \Rightarrow \mathbf{n}_k^m$ we have

$$(7) \quad \Pr\{\mathbf{n} \Rightarrow \mathbf{n}_k^m\} = \Pr\{\mathbf{n} \rightarrowtail \mathbf{n}_k\}\Pr\{\mathbf{n}_k \rightarrowtail \mathbf{n}_k^m\} = \frac{n_k(w_m + n_m)}{n(w + n - 1)}.$$

(7) ensures that this is the transition probability of a homogeneous, aperiodic Markov chain for which we can single out the equilibrium distribution (using the conditions of detailed balance). (7) is the basis of the unary transition matrix. Transitions of a greater degree are mere enlargements of (7) (Costantini and Garibaldi, forthcoming c). For example, taking a binary transition into account, $\Pr\{\mathbf{n} \Rightarrow \mathbf{n}_{kh}^{ml}\}$ is reached in exactly the same way as that shown by (5) and (6), considering two destructions in the mth and lth cells and two creations in the kth and hth cells.

6 Conclusion

According to Popper "propensity is [...] a somewhat abstract kind of physical property; nevertheless it is a real physical property" (Popper 1967, p. 33) . Also the creation propensity we have characterized can be interpreted in this way. That is, this propensity can be considered as a physical property of the experimental set-up composed of individuals and categories, in physics of cells and particles. From this point of view the creation propensity acts like a probabilistic force. More exactly, like a relation between cells filled with particles and particles to be created. This relation can also be seen as a probabilistic field, that can give rise to repulsion and attraction. Obviously, it is a relation that takes place not in real space but in the μ-space, in general, in the space of categories. If the system is void, each cell has the same propensity to accommodate a particle, or the probabilistic field is the same for all void cells. On the contrary, if some particles are in a cell the field is proportional to the particles already created in it. For any number of particles and any cell, the intensity of the field generated by the particles in the cell is proportional to their number. The proportionality factor is fixed by the parameter c of (4).

Perhaps the best example is given by the collision of two particles ruled by a binary transition. Such a physical interaction operates as follows. First, it destroys at random, that is all particles have the same propensity to be destroyed, irrespective of the cell in which they are. As a consequence the cell with more particles has a great propensity to suffer a destruction. Afterwards, creations act in the direction of equilibrium, that is the creation of a particle in a state is more probable the farther the state is from equilibrium. Whenever a collision takes place, the transition of the particles drives the system towards equilibrium. It is worth noting that equilibrium is a dynamical state generated by the transition propensity.

Seeing the matter in this way, no maximum entropy principle is considered. This is because we have used probabilistic conditions from the beginning.. Equilibrium is determined by the regularity and propensity conditions. The astonishing simplicity of the derivation we have given—it can easily be generalized to Canonical and Grandcanonical cases—ensues by approaching the matter probabilistically from the beginning.

As a final remark we note that for large systems the average occupation number of a cell resulting from the probability we have considered is

$$\chi_k = \frac{1}{\exp\left[\frac{(\epsilon_k - \mu)}{kT}\right] - c}$$

that generalizes the physical cases to any value of c also in the case of "parastatistics" (Costantini and Garibaldi 1991). By putting the value of c equal to $-1, 0$, and $+1$, we transform probabilities into propensities obtaining the average occupation number of the physical cases.

References

Brillouin, L. (1927) Comparison des différentes statistiques appliqué aux problème des quanta. *Annales de Physique Paris,* VII, 315–331.

Costantini, D. and Garibaldi, U. (1989) Classical and quantum statistics as finite random processes. *Foundations of Physics,* 19, 743–754.

Costantini, D. and Garibaldi, U. (1991) Una formulazione probabilistica del principio di esclusione di Pauli nel contesto delle inferenze predittive. *Statistica,* LI, 21–34.

Costantini, D. and Garibaldi, U. (1997/1998) A probabilistic foundation of elementary particle statistics. Part I. *Stud. Hist. Phil. of Modern Physics,* 28, pp. 483-506; A probabilistic foundation of elementary particle statistics.. Part II, *Stud. Hist. Phil. of Modern Physics,* 29, (1998) 37–59.

Costantini, D. and Garibaldi, U. (forthcoming a) New characterizations for equilibrium models in population genetics.

Costantini, D. and Garibaldi, U. (forthcoming b) A probability theory for macroeconomic modelling.

Costantini, D. and Garibaldi, U. (forthcoming c) A purely probabilistic representation for the dynamics of a gas of particles. *Foundations of Physics.*

Dirac, P. A. (1958) *The Principles of Quantum Mechanics.* Oxford: Clarendon Press.

Galavotti, M. C. (1999) Some remarks on objective chance (Ramsey, Popper and Campbell). In Dalla Chiara et al. (Eds.) *Language, Quantum, Music.* Dordrecht-Boston: Kluwer, 73–82.

Popper, K. R. (1957) The propensity interpretation of the calculus of probability and the quantum theory. In S. Körner, (Ed.) *Observation and Interpretation.* London: Butterworths Scient. Publ., 65–70.

Popper, K. R. (1959) The propensity interpretation of probability. *British Journal for the Philosophy of Science,* X, 25–42.

Popper, K. R. (1967) Quantum mechanics without 'The Observer'. In M. Bunge (Ed.), *Quantum Theory and Reality.* New York: Springer, 7–42.

Suppes, P. (1974) Popper's analysis of probability in quantum mechanics. In P. Schlipp (Ed.), *The Philosophy of Karl Popper.* La Salle IL: Open Court, 760–774.

Suppes, P. (1987) Propensity representations of probability. *Erkenntnis,* 26, 335–358.

14

Physical Quantum States and the Meaning of Probability

MICHEL PATY

Centre National de la Recherche Scientifique
University of Paris F. D. Diderot

1 Introduction

It is intended, in what follows, to perform an epistemological analysis of probability statements in quantum physics, in the perspective of an *extension of meaning* of the *notions of physical state* and of *physical quantity*, restricted up to now to numerically valued forms, to more complex mathematical expressions of the type that are used in the quantum theoretical formalism, i.e., state functions as coherent superposition of basis (eigen) state vectors of a Hilbert space and, for quantities or magnitudes, matrix or linear hermitian operators acting on the state functions.

In the usual interpretation of the "quantum formalism", these forms are considered as purely mathematical, their physical meaning being considered as given from *interpretation rules* relative to *measurements* performed on "quantum systems" with apparatuses obeying the laws of classical physics. The proposed extension for the concept of physical state and quantity would allow to claim that, contrary to the received interpretation of quantum mechanics, the *state function represents directly the physical system* in the considered state (instead of being viewed as a "catalogue of our knowledge of the system"), and that *the theoretical quantum quantities* (the "observables", as they are currently called) *represent physical magnitudes*, the dynamical

Stochastic Causality.
Maria Carla Galavotti, Patrick Suppes and Domenico Costantini (eds.).
Copyright © 2001, CSLI Publications.

variables, that *are properties of the system*. Such a view would considerably simplify the "interpretation problems" of quantum physics, for one could henceforth speak, for this domain, not only of phenomena related with our observation of them but, in the same way as in the other areas of physics, of physical systems having properties, i.e. *objects*, standing independently of our knowledge of them and that are fully described by quantum theory.

We have argued elsewhere about various aspects of this unorthodox point of view, concerning the physical as well as the philosophical aspects of interpretation, the first one including the question of the physical meaning of theoretical (mathematically expressed) quantities, and also the problem of the quantum to classical relationships (Paty, 1999, 2000a, 2000b, forthcoming b). Such a *direct physical interpretation of the quantum variables* seems actually to correspond to the implicit conception of the quantum physicists at work, not only at the theoretical level, but also when considering the physical implications of it as manifested through experiments. This implicit conception, in our view, can rightly be made explicit, being justified with sound arguments, considering the quantum phenomena in their variety, from the "simplest" ones (such as quantum interferences) up to the most elaborated ones (taken in subatomic physics and quantum theory of gauge fields). It might therefore appear nowadays as the most natural interpretation of the quantum formalism.

We shall concentrate here on the consistency of this "direct interpretation" conception with our understanding of probabilities in quantum physics. As a matter of fact, a questioning of the peculiar character of probabilities at work in quantum physics (through the concept of "probability amplitude") leads, as we shall see, to clarifications and distinctions concerning probability and statistics, theoretical (quantum) quantities and (classical) measured ones; such clarifications call to the forefront a fundamental property attached originally to the notion of *physical quantity*, that of being *relational*. These clarifications are fully consistent with the proposed direct interpretation for the state function and the quantum theoretical variables (as operators) in terms of a *theoretical description* (or representation) *of physical states and of their physical properties*.

We shall sketch briefly how the concept of probability allowed to conceptually penetrate the world of *atomic* (and henceforth *quantum*) *physics*, by concomitantly undergoing changes of meaning and afterward of function. From a pure mathematical function used in auxiliary reasoning, it became "physically interpreted" as frequencies of occurrence of events, and turned in this way into a fecund heuristic tool that helped revealing specific features of quantum phenomena and systems. It happened further, in the course of the evolution of ideas in quantum physics, to structure progressively the physicists' comprehension of the peculiarity of the new quantum domain, and at the same time it was merged into the problems and ambiguities of the "interpretation" of quantum mechanics, being tightly connected with the "measurement problem", as well as with the quantum to classical relationship.

It is this problem of interpretation that we have in mind, and we shall try, from our analysis, to clarify the exact *meaning of probability statements* in quantum physics. Meaning, or meanings ?... For, we shall establish conceptual distinctions between probability in the theory and probability from results of experiment : two different meanings of probability as implied in quantum physics, respectively a *quantum* and a *classical* one, a *relational* and a *statistical* one, that the "probability interpretation" taken together with the "measurement rule" had merged into one another, which had led to some confusion.

2 Probability as a Tool to Explore Quantum Phenomena and the Classical Physical Meaning

Probabilities play in quantum physics a much more fundamental and deeper role than in classical physics. Actually, probabilities entered physics in a quite classical way, that is to say, in the ancillary position of a mere pragmatical means, and such was still their status when they happened to be taken as a powerful and indispensable tool in the first physical investigations of the microphysical (atomic) world (Boltzmann 1896/1898, Gibbs 1903). But afterwards, when this atomic world revealed itself qualified in a decidedly nonclassical way as a world of *quantum phenomena* and supposedly of *quantum entities* responsible for these phenomena, probability began showing more than indispensable : it entered the very foundations of the description of this world. Let us recall the gross features of this status transformation of probability when its use was moved from classical to quantum physics, which one could summarize as: *from distributions to structuration*. Such a change would correspond to an underlying modification in our understanding of the basic notions used in physics, such as that of quantity: we shall retain, as a great epistemological lesson of these developments, the move of our notion of *physical quantities, from measurement to relationship*. Further on we shall make these statements more explicit.

For the atomic domain, which escaped the possibility of direct observation, statistical mechanics happened to be, after chemical analysis, the most natural tool to explore it, in a conceptual continuation of statistical mechanics along Ludwig Boltzmann's path, as Max Planck did with the study of radiation emitted and absorbed by atomic matter in a "black-body" (heated closed cavity at thermal equilibrium). The indirect exploration thus opened would reveal for the new invisible area, notwithstanding the methodological continuity, characteristics escaping any reduction to the usual conceptual schemes of "classical" physics. As a matter of fact, Planck had to perform empirically some modifications to the usual mode of counting complexions *à la* Boltzmann for elements of radiation energy, which led him to discover the discontinuity of energy exchanges between atomic matter and radiation[1].

The first moment of the introduction of probability in what would become quantum physics, was that of mathematical treatment by Planck of the energy distribution of radiation in analogy with statistical mechanics: Planck adapted to his problem the method of statistical physics developed by Boltzmann for atomic gases: he transferred a procedure adequate to a discontinuous sample (an atomic gas distributed in a volume of phase space) to the case of a continuous one (that of radiation energy distribution). Clearly, the probability considered in both cases had not the same meaning from the physical point of view. With Boltzmann's statistical physics of gases, counting particles and states leads to a combinatorics of numbers of particles for given volume elements of phase space, giving probabilities from integer values that had somehow a physical significance. Planck was well aware that the probability distributions he attained at for radiation in a similar way had not the same physical meaning, and considered the calculated probabilities in his problem as purely mathematical intermediaries in the reasoning. His theoretical treatment of radiation, although fruitful with regard to the result obtained, looked artificial and had no justification if not only a pragmatical one. The procedure used to obtain it was rather obscure (Jammer 1966, Kuhn 1978, Klein 1962, Kastler 1983).

Some time later, Einstein radicalized the idea of the discontinuity of energy, making of it and of Planck's quantum of action a general property of radiation[2] as well as of atomic matter[3] themselves, and not only of their mutual exchanges, arriving soon at the conclusion that this property was irreducible to classical physics and implied fundamental modifications in both the electromagnetic theory and the mechanics of bodies at atomic level. To get at this conviction (as soon as 1906), he made use of a physical reasoning he had set up (already in 1903) about the physical meaning that was to be given to the probability implied in the theoretical calculations. The probability involved in the combinatory of complexions of energy cells was purely

---

[1] In his one hundred years ago celebrated publication, Planck (1900), although he would try for many years to restore some fundamental continuity. Planck's energy relation is $\Delta E = nh\nu$ (ΔE: radiation energy absorbed or emitted, ν: radiation frequency, h: Planck's constant of action, n: integer number).

[2] First in Einstein (1905) and in Einstein (1906), where fluctuation calculations around statistical mean values led him to diagnose in radiation a combination of wave interference and corpuscular contributions.

[3] Einstein published his first paper on specific heats in 1907 (Einstein 1907). See also Einstein (1912).

mathematical, like counting balls in urns, which would look rather improper if it were considered physically for radiation, considered as continuous according to the electromagnetic theory (as already emphasized above).

For physical phenomena and quantities, Einstein thought necessary to *reinterpret physically* the mathematical probability: he gave to that one present in Boltzmann's entropy formula (which he would use to call "Boltzmann's principle"[4]), the meaning of a frequency in time (for a system to come to the same physical state). Frequency in time is a physical quantity that can be empirically determined, with a mean value and fluctuations around the mean value (Einstein 1903). Such fluctuations were not considered with pure combinatoric calculations although they effectively characterize the distributions of quantities relative to physical phenomena.

From this *physical reinterpretation* of a *mathematical quantity* implied by idealized processes of combinatorics, calculating fluctuations became Einstein's favoured exploration instrument for the atomic and quantum domain, which led him to his major contributions in this field from 1905 to 1925: evidence for molecular motion, light quanta as shown in particular in the photoelectric effect, both in 1905[5]; some dual wave-corpuscle aspect of radiation in energy distributions, from 1906 to 1909; extension of Planck's quantum of action to the atomic structure itself and specific heats, in 1907 up to 1911; first synthetic theory of quanta with evidence that light quanta carry momentum, in 1916-1917; statistical specific properties of radiation and similarities with monoatomic gases, i.e., Bose-Einstein statistics for indistinguishable quantum particles, in 1924-1925[6], related with the generalization of the wave-particle duality for any elementary material system as formulated by Louis de Broglie in 1923 (de Broglie 1924).

Calculating fluctuations in the statistical distributions of significant dynamical quantities was a powerful tool to getting a knowledge of characteristics of quantum phenomena. As a matter of fact, the properties of the quantum domain have been mostly expressed in a statistical-probabilistic way. For instance, the first indices for some kind of particle-wave dualism obtained as soon as in 1906 had the form of a juxtaposition, in an energy fluctuation formula, of a statistical mechanics type term and of an interference one.

Fluctuations, that were the mark of a physical meaning for probability distributions, revealed actually in Einstein's hands properties of the new ra-

---

[4] The Boltzmann's formula is: $S = kLogW$ (S: entropy, W: probability of the state, k: Boltzmann's constant (Boltzmann 1896-1898).
[5] The light quantum is characterized by its energy (E)-frequency (v) relation : $E = hv$.
[6] Respectively, Einstein (1905a and b; 1906, 1909, 1907, 1911, 1916-1917; 1924, 1925). See Kuhn (1978), Darrigol (1988, 1991), Paty (forthcoming a).

diation and atomic domain of phenomena (the quantum domain) that were decidedly not classical ones. Among such properties were the energy discontinuity of radiation and of atomic levels, dual wave-particle aspects for light, and quantum statistical behaviour (for bosons, revealed also for fermions with Dirac) referred to the indistiguishability of identical quantum systems. All three appeared to be fundamental properties of the quantum systems, interconnected with each other. The latter, in particular, entailed powerful consequences that have all been verified afterwards (explanation of the periodical table of the chemical elements, constitution of degenerate stars, Bose-Einstein condensation of many identical atoms falling into the same "zero energy point" ground state, ...) (Paty 1999, forthcoming b).

It must be noted, incidentally, that probability as statistics entered also early in quantum physics from another, independent, way : the law of radioactive decays formulated in 1903 by Ernest Rutherford (1952-1965, vol 1) and Frederic Soddy, expressing a constant rate of disintegrations in time, which corresponds to the independence of successive events in radioactivity (Amaldi 1979). This law was extended to atomic transitions, from Bohr's and Sommerfeld's 1913-1916 atomic model to Einstein's first (semi-classical) quantum theory of 1916[7].

The consideration, in Einstein's 1916-1917 quantum papers, that the light energy quantum has a momentum[8], which entailed its full particle property, came out from the condition of equilibrium of a statistical ensemble of quanta of radiation and of atomic states emitting and absorbing this radiation with amplitudes of transitions ruled by statistical laws. The relative frequencies of the states could be obtained from thermodynamical considerations or from "Boltzmann's principle", and the transition probability between two states was expressed in the same way as that of radioactive decay[9]. Einstein obtained from it Bohr's quantum condition for transitions between atomic levels[10] and a derivation of Planck's radiation formula. In this first theory of quanta, yet a semi-classical one, the *amplitudes* of transitions between atomic levels were characteristics of the time probability law of the process, given by statistical distributions. (At this time they were given from experi-

---

[7] See Bohr's 1913 paper in (Bohr 1972), Einstein (1916-1917) and Paty (1988 and forthcoming a).

[8] $p = \dfrac{h}{\lambda}$ (p : impulsion; λ : wave length). Einstein (1916-1917).

[9] $dW = A_m^n dt$. The transition amplitude A_m^n has the same paper as a radioactive constant, characteristic of a given radioactive substance.

[10] Bohr's condition is: $|E_m - E_n| = h\nu_{mn}$ (E_m and E_n : energies of atomic levels m and n; ν_{mn} : frequency of the radiation emitted or absorbed between these levels).

mental data. They would be calculable theoretically only with quantum field theory, shortly after the quantum mechanics formulation was obtained (See, for example, Born and Heisenberg 1928)).

The evidence for attributing radiation (defined by its frequency and wave length) both an energy and a momentum (property of a particle), seemed to show that atomic processes were defined at the level of individuals. But, as Einstein observed in his conclusion, by which he considered that a "proper quantum theory of radiation appear[s] almost unavoidable", such a full theoretical understanding was not yet achieved : "The weakness of the theory lies on the one hand in the fact that it does not get us any closer to making the connection with a wave theory; on the other, that it leaves the duration and direction of the elementary process to "chance"" (Einstein 1916-1917, Engl. trans. p. 76).

Referring probability to the law of chance was expressing its classical nature (with its "subjective", laplacean, interpretation (Laplace 1814)). As a matter of fact, the use of probability in quantum physics up to then had been able to shed light "from outside", so to speak (i.e., from the concepts for classical phenomena) on genuine characteristics of quantum phenomena, irreducible to classical ones. To help going further into the quantum domain, probability would have to uncover a radical change in its function and meaning, as the problem at stake was to formulate a *proper quantum theory*, as Einstein said. This would not be unthinkable, since one would always have the possibility to afford probability, which as such is a purely mathematical concept, a different meaning for its use in physics[11]. The identification of a probability with a frequency was a choice justified in classical physics. Quantum physics may lead to give privilege to another kind of physical interpretation. Although the problem was not put in these terms at the time, we point out this alternative as a possible reading of quantum mechanics.

3   Probability for Individuals: Changes in the Meaning

One of the major problems of the physics of quanta was set from that time onwards: how to reconcile in a "proper quantum theory" a "probabilistic determination" of the properties of physical systems with the conception that these systems are made of individual entities?

The Bohr, Kramer and Slater's episode might be viewed here as a symptom of this problem before quantum mechanics was formulated. After Compton's experiment had confirmed by observation, in 1923, the momen-

---

[11] On the mathematical theory of probability, see Kolmogorov (1933). For discussions about the meaning attached to probability in physics, see, for instance, Popper (1957, 1982, 1990) and Bunge (1985).

tum of quantum radiation (theoretically derived by Einstein), i.e. its corpuscular character, from a collision process such as the extraction of an atomic electron by an X-ray of a given wave-length and direction impinging on it (Compton effect (Compton 1923))[12]. It was just a matter of writing the momentum-energy balance in the reaction for particles, and comparing with the values obtained from the detection of the emitted electron and of the scattered photon. However, Niels Bohr, Hendrik Kramer and John Slater put doubt on the conclusion, in an attempt to maintain continuous energy exchanges inside the atom that they wanted to conciliate with discontinuity in quantum phenomena (Bohn, Kramer and Slater 1924)[13]. They argued that momentum-energy conservation might not hold at the atomic level for individuals but might be only statistically verified. The experiment performed in 1925 by Geiger and Bothe, setting evidence for an individual correlation between the electron and the X-ray photon simultaneously emitted was an unambiguous proof of the particle property of radiation (Geiger and Bothe 1924, 1925). It was at the same time a proof of the "individual reality" of light quanta, as Einstein stated in his own exposition of the result (Einstein 1926)[14].

Significantly enough, it was this individual physical reality of quantum systems that became thereafter Einstein's main concern about quantum mechanics[15]. How far was this theory able to describe individual physical systems? The answer was not a priori obvious, neither for the opponents to the standard Copenhagen interpretation such as Einstein, nor for its proponents, such as Bohr or Born, and others. The fact that probabilities had an important paper did not forbid the possibility of getting at some description, even indirect, of individual quantum processes and systems. The problem with quantum physics was that such individual systems were exhibiting unusual properties such as the wave-particle dualism and types of correlations showing up in scattering, in interferences, in "quantum statistical behaviour" and, finally, in non-local separability of subsystems, all of them appearing to have something to do with a non-local character.

A decisive aspect in these circumstances was qualified when one realized, in 1925-1926, that the quantum statistical behaviour of radiation (that of symmetric indistinguishable entities, bosons, with respect to their mutual exchange) was the deep reason that justified Planck's unusual counting of

---

[12] The experiment was done with low atomic number elements.

[13] In this theory, the Compton scattering of the electron would be a continuous process in which all atomic electrons of the scattering body take part, only emitted electrons being individuals and being submitted to statistical laws.

[14] See also Einstein's correspondence of the time with Langevin, and others, mentioned in Paty (forthcoming a).

[15] For his further inquiries, and in particular the EPR argument, see Paty (forthcoming a).

radiation energy distributions by substituting combinations to permutations, obtaining as a consequence discontinuous energy exchanges (although he was not aware of it). He did it actually with a purely pragmatic purpose, that of recovering the observed spectrum of frequencies for radiation in the black body. This unusual way of counting had been noticed and analyzed by Ladislaw Natanson and by Paul and Tania Ehrenfest already in 1911-1912 (Natanson 1911; Ehrenfest 1911; Kastler 1983; Darrigol 1988, 1991), but it had to wait twelve more years to be fully acknowledged. This deep origin (which remained hidden underground for a long time) makes us aware that the quantum statistical or probabilistic dependence was, so to speak, co-natural to the quantum of action which is the mark of the discontinuity of energy and, more generally, of the quantum specificity.

And then goes on the quantum physics story (and history). Based on all the quantum properties known by then, whose deep reason, or "essential feature", can be identified as *quantum statistics* or indistinguishability of quantum systems, quantum mechanics was built, with probability still being of help (although not reinterpreted), as a powerful tool, in the following sense : quantum mechanics was leading to probability statements that were confirmed by experiments, in statistical distributions. The theory, quantum mechanics, took the form of a mathematical-theoretical machinery aimed at the description (by then thought as an indirect one) of the states of a physical system, based on mathematically expressed quantities such as the *state function* (vector of a Hilbert space, denoted ψ) ruled by the (generally considered as "formal") *superposition principle* (by which, ψ actually appears as a *phase coherent* vector superposition of basis eigenstates), and the theoretical quantities, or quantum variables, usually called "observables", were represented by linear hermitian non-commuting operators (a "formal" property again) (Dirac 1930; Neumann 1932)[16].

At this stage, the theoretical structure was considered to be a mathematical one, whose (mathematical) entities had the function of *bearing the relationships* that were characteristic of the physical quantum phenomena and systems. But they were not considered by physicists, at least when trying to elucidate their relations to physical contents, as *physical quantities*, for the reasons many times discussed in the quantum debate about interpretation (Bohr 1957; Einstein and Born 1969). Physical contents were thought to come only from experiments, performed with macroscopic devices, whose results were stated in terms of classical quantities statistically distributed. These classical quantities showed being submitted to restrictions (or to "conditions of use") such as Heisenberg inequalities, which transcribe their being related to quantum systems.

---

[16] For an historical recollection, see Jammer (1966, 1974).

4 Physical Meanings: From Measurement to Relationship

By having been able to reveal, from a classical approach, unclassical features, probability underwent a change in its function and, possibly in its nature, with respect to physics and to physical theory. And physical theory is generally made of physical quantities. Even if the nature of such quantities for quantum physics has remained for a long time unclear, the evolution just mentioned might be evaluated with regard to something of physical quantities. Let us state that this evolution has gone along not so much from classical measurements to quantum measurements, as it is usually presented, but *from a thought of measurement* (whatever the meaning of "measurement" was thought to be) to a thought about *relationships*. And, to introduce already what I have in mind and shall emphasize afterwards, this is precisely why probability got more importance in quantum than in classical physics, for whereas relationship is *internal* to the descriptive theoretical and conceptual scheme, measurement is only *external* to it.

Relationing (relationing quantities one to the others) corresponds to the essential *function* of the *quantities* that express physical concepts. In classical physics, the relationships of quantities have taken the form of causality in the "Newtonian" sense of the differential dynamical law. The causality law entailed the requirement of precision in theoretical as well as in experimental determination, and this is how probability entered classical physics, by two ways. By the way of *measurement* of physical quantities, with the theory of errors, which was the means to counterbalance the *lack of experimental precision*, on the one hand. And, on the other hand, by the way of *observation* of quantities that were assigned to be the average of other ones, in order to counterbalance *uncomplete knowledge* as in statistical mechanics. With these two statistical-probability "repairs", the ideal of causality was, so to speak, recovered (in the vein of Laplace's philosophy of probability (Laplace 1814)).

But at the same time, by being considered as "formal", they were cleaned out of any physical content, because their (formal) properties appeared contrary to what was usually understood as physical properties, which ought to be expressed (so one thought) through numerically valued variables and functions. If the quantum theoretical quantities were only *mathematical*, the *physical content* of the theory was thought as being given through the *interpretation rules* relating the theoretical formalism and the observational data.

Considering quantum theory in this manner would actually be to make of it a phantom relating two disconnected orders of things: the formal and mathematical one and the empirical data given in experiments. Strictly speaking, the physical theory would reduce to the quantum interpretation rules connecting, in a purely conventional way, the mathematical or formal and the empirical. Up to that circumstance, physical theory used to be considered as a theoretical structure of concepts having physical content. One may ask whether separating in two distinct moments and functions *probability statements* and *measurement* would not help to recover some scheme of this kind for the physical theory of the quantum domain.

5 Probability Interpretation and the Measurement Rule. How to Escape the Procuste's Bed?

Consider these two rules of the quantum formalism: the *probabilistic or statistical interpretation* of the state function or state vector, and the *measurement or reduction rule* of the state function to one of its components in the observation process. These two rules have been tightly connected in the formulation of quantum mechanics as an axiomatic theory (von Neumann 1932; Dirac 1930), due to the fact that measurements yield statistical distributions for the various states of the system and for the corresponding values of the compatible variables[17] characterizing these states. In effect, the measurement interpretation *rule* had been formulated in such a way as to seal quantum mechanics as a closed system, as if it were a *principle* for quantum physics, when it is merely a pragmatic statement, a recipe for use, imposed by the necessity in which we are to get information about quantum physical systems through classical observation and measurement devices. This Procuste's bed condition can in no case be invoked as defining quantum systems, since these systems cannot be reduced to classical properties. It is obviously a human (macroscopic) observer's limitation with respect to the entities, whatever they are, of the quantum domain of physical reality.

We might actually consider the two statements separately, and the first one (the probabilistic interpretation) had indeed been formulated independently from the second, and previously to it by Max Born (1926). Some distance must be taken from the historical circumstances of the edification of quantum mechanics, when the new pieces of the quantum puzzle seemed to organize themselves in a so incredibly consistent and powerful way as to inspire its inventors the conviction that the theory was already (in 1927) complete and definitive (Born and Heisenberg 1928). Now that we have no doubt that quantum physics is a sound piece of knowledge about a large part of world phenomena, we may allow ourselves to loosen the elements of the logical construction and think afresh the meanings of these statements. Actually, the two rules are by no means tied together in essence. Let us first consider separately the meaning of the first one, up to the point when we shall

---

[17] i.e., whose operators mutually commute.

need the second.

The probabilistic interpretation of the state function is obviously one of the main foundational statements of quantum mechanics as a physical theory: it defines the correspondence between a chosen mathematical quantity and a physical concept. Or, in other words, it defines a physical quantity with the help of a mathematical expression, and this is, indeed, how physical theory usually proceeds. In this respect, we may consider that such a quantum physical construction of concept does not differ from what one has been used to do in classical physics (including relativity theory), a process that is responsible for the "success of mathematics in physics", which is the counterpart of the mathematization of this science, entangled with its conceptual edification.

If we take this path, we meet the question of the *physical meaning* or *content* of a *state probability* in quantum physics. Actually, probability is only one of the steps of the interpretation, so to speak the last (or the synthetical) one and, significantly enough, physicists have been led to identify, as conceptually previous to it, an *amplitude*, which they called *amplitude of probability*. Although such an expression sounds somewhat non realistic (a probability could hardly be some kind of physical object propagating through space), it expressed at the same time a *necessity* and an *impossibility*. The impossibility was clearly realized as soon as Schrödinger's equation had been established (Schrödinger 1926). The "wave equation" adequate to the description of a quantum system (for instance, an hydrogen atom) is about something else than a physical wave, whose propagation in space would not hold the quantum properties of the considered system, which would be spread away[18]. Nevertheless, if there is no wave for the wave equation, there is *something* having properties *mathematically* (or *formally*) *analogous to the amplitude of a wave*, and this is the *state function* itself. Entered mathematically in the equation for the physical system obtained through an hamiltonian formalism, the state function must be given back a physical content from the successful application of the mathematical formalism to the physical properties of the system.

The physical content of the state function and of the related dynamical quantities expressed mathematically lies in the relationships they ascribe to the corresponding physical quantities that were considered as such from the start (those being measurable), i.e., the basis state functions as the solutions of the state equation (eigenfunctions), and the eigenvalues of the "observables" as operators). In particular, the overall ("mathematical") state function works, in this respect, in a fashion similar to that of an *amplitude* of wave in

---

[18] I refer to the debate that took place shortly after Schrödinger's derivation of his equation, in 1926 (Schrödinger 1926; and see Jammer 1974; Paty 1993b).

wave theory : in particular, as a coherent linear superposition of basis states functions, it entails interferences between these states, that are indeed observed. The mathematical overall state function (ψ) can therefore be itself interpreted as a physical *amplitude*, in the sense of giving rise to interference phenomena whose intensity is given by the squared modulus of it. It thus is the *amplitude of something*, but *of what*, if not of a wave ? But although suggestive, the wave analogy is restrictive, and the state function of a quantum system is actually much richer of physical content than that of a mere wave amplitude, as it holds all the specific (non classical) properties of quantum systems, such as, for instance, interference of a single quantum state with itself, non local separability of correlated sub-states, Bose-Einstein condensation for identical bosons, etc. (Paty forthcoming b).

We have obtained, at this stage, an important conceptual result about the physical meaning of the "mathematical" quantum variables or dynamical quantities: they express *relations* that are characteristic of quantum phenomena and are revealed in observation and measurement. These ascertained relations hold on *quantities prepared for measurement* with classical devices and which we can consider as the classical projections of "quantum quantities" characteristic of the quantum systems. Such *relations* are actually *deduced from statistical results* obtained for these classical quantities which, taken together, carry in a way or another the specific quantum character of the intitial system before its measurement. Conceptually and theoretically one has to reconstitute this quantum characteristics that shows in the relationship between the classical quantities projected from the quantum ones, and given to us from their statistical distributions.

6 Theoretical and Empirical Probabilities

Clearly, what precedes suggests the need, at this stage, of conceptual distinctions between two different effective uses of probability in quantum physics : on one side, a *theoretical* definition of *probability* in the quantum description of physical systems, even of *individual* systems; on the other side, an *empirical* acception, where it refers to the statistical results of experiments. In the first sense, probability properly speaking, expressed as a mathematical function, is afforded a *theoretical physical meaning*, enrooted in the specific, physically elaborated, concept of "probability amplitude" (whose denomination, historically determined, remains somewhat ambiguous); it is *quantum theoretical and relational*. In the second sense, it is given a purely statistical meaning in the same way as when it is used in classical physics, in statistical mechanics for instance: it has no more a theoretical (and quantum) function, but a practical one, that of expressing results of experiments and of measurements, performed on quantum systems, in terms of classical quantities[19].

"Probability amplitude" is a bizarre expression for a concept or a quantity in physics. This óddness may have been the signature of its impossibility to be physically thought in a direct way. Made on the mold of "wave amplitude", which it could not be, it does not either correspond semantically to something analogous: "probability" being the squared modulus of amplitude, its analogous for classical waves would be "intensity". But "intensity amplitude" would be tautological and is not, indeed, in use for waves. The exact analogous to "wave amplitude" would be "quantum state amplitude", which is effectively used also by to-day quantum physicists. As for them, the "founding fathers" of quantum mechanics preferred to speak in this sense of "state function" or "state vector", which they thought as mathematical quantities, and not as physical ones. As an effect, they forged this queer expression, "probability amplitude", when they realized that the solution of a wave equation for a quantum system could not be the amplitude of a wave, as recalled before.

Although of an obscure meaning literally speaking, the expression "probability amplitude" used to designate the state function must be given the credit of referring it, even if reluctantly, to something physical (an amplitude) and at the same time to its correspondents on the side of usual (classical) quantities (probability related to statistics). Note that the word "probability" gets a significant position here, as it could not be substituted by "statistical" (for, what would be the meaning of an expression such as "statistical amplitude"? it would be not only queer, but nonsense), notwithstanding the lack of precision already diagnosed among the founding fathers concerning probability and statistics. "Probability amplitude" is indeed a quaint and at the same time a significantly penetrant concept as, by juxtaposing two terms so much foreigner one to the other, it gives (quantum) probability a *physical content* and provides a mathematical (state) function with a *precise theoretical meaning* related through a straightforward correspondence with empirically determinable quantities. Such was the insight, may we think from our proposed point of view of a "direct interpretation", but it was by then inhibited by the compelling orthodoxy... Even scientific terminology is affected by historical contingency. Let us keep the queer expression, "probability amplitude", as culturally useful, reminding the uncertain paths of the discovery, of how one has come to know what was unknown.

---

[19] It is impossible to do justice to those who have helped in trying to clarify the uses and meanings of probability in quantum physics. About the analyses and possible interpretations of probability in quantum physics, I am indebted (non exhaustively) to: Reichenbach (1944, 1978), Popper (1957, 1982, 1990), Suppes (1961, 1963, 1970), Schushurin (1977), Bunge (1985); and see Mugur-Schächter (1977), Paty (1990). In the strict Copenhagen sense, see, for instance, Rosenfeld (1974).

The theoretical probability (actually, its "amplitude", amplitude of something, whatever it be) is reconstituted from the set of values of measured quantities with their corresponding probabilities. This quantum state amplitude, or state function, is therefore identified as the true *source*, or to be more exact, as the representent (or theoretical description) of the true *source of the ascertained physical relationships,* this source quantum theory is aimed at which is in right to be called physical: the state of the system, beyond its projections. The *relational disposition* of the state function (with its related quantum dynamical variables) given by its mathematical form, appears unambiguously as the theoretical counterpart of the physical properties that have been registered, and can therefore be endowed with a straightforward physical meaning: the *state function* may be taken as the very *theoretical expression of its physical content.* By "the very expression", I mean it as the full and most economical one. This means that the "mathematical quantities" (as physicists were used to think of them) of the quantum formalism should be considered as *physical* ones, but with two differences with respect to what is generally understood as *physical quantities* in the classical sense: they are expressed by mathematical quantities more complex than simple numerically valued ones, and they are only indirectly given (by rational reconstitution) through experiments of a classical type (with statistical distributions of classical quantities).

At this stage, a conceptual distinction needs to be emphasized between two different effective uses of probability in quantum physics: on one side, a *theoretical* definition of *probability* in the quantum description of physical systems, even of *individual* systems; on the other side, an *empirical* acceptation, where it refers to the statistical results of experiments. In the first sense, probability properly speaking, expressed as a mathematical function and used in theoretical calculations, is afforded a *theoretical physical meaning*, enrooted in the specific, physically elaborated, concept of "probability amplitude" (despite the ambiguity of this denomination, historically determined); it is *quantum theoretical and relational*. In the second sense, probability is obtained from the statistical results of measurements. These measurements are performed *on quantum systems* but *in terms of classical quantities*. Probability obtained in that way have a purely statistical meaning, referring to the distributions of values of classical states and quantities as, for instance, in classical statistical mechanics; with respect to quantum systems, it has no theoretical function, but only a practical one.

Through the measurement process of the quantum system, the theoretical, quantum, relational probability is put in correspondence with the empirical, classical, statistical one, the quantum proper theoretical description is confronted to the response of experiment. It is in this way that classical apparatuses of our macrocosm have been opening a window on the microcosm of the quantum world. The clear distinction which we have tried to establish between a physical theoretical description and the corresponding empirical data, allows with full right to speak of a proper quantum world which can be *thought independently of measurement*, i.e., of our interaction with it. By "independently" it is meant that it stands on its own reality and soundness. As in other fields of knowledge, and particularly as in classical physics, it is known to us through the symbolic and conceptual, theoretical, representations that we make of it, and these representations are fed with the data of experiments.

The conviction that it is possible to conceive the reality of the quantum domain, by affording *a full direct physical meaning* to its theoretical representation in the way just exposed seems therefore to be rather firmly sustained. To get soundness, two related conditions have been particularly operative: making a distinction between two different physical meanings of probabilities as used in quantum physics, one quantum theoretical (relational and mathematical) and one empirical (probability in a statistical sense), and disconnecting, from a fundamental point of view, the "probabilistic interpretation" rule of quantum theory from the "measurement" or "reduction" one.

Indeed, the *connection* between these two *probabilities*, the *quantum theoretical and relational one* and the *classical empirical statistical one* might be viewed as the remaining most fundamental *interpretation problem* of quantum physics, intending this time not so much the physical as the philosophical interpretation, because it points directly at the modalities of knowledge. One might refer, up to some extent, such a distinction to the respective roles of understanding and perception, rational elaboration and data acquisition. Let us content ourselves here by concluding with the simple remark that connecting is not identifying. It seems that, with the case just discussed, connecting opens intelligibility anew, whereas, on the contrary, identifying is limiting and shuts down to obscurity.

References

Amaldi, E. (1979) Radioactivity, a pragmatic pillar of probabilistic conceptions. In G. Toraldo di Francia (Ed.), 1-28.

Bohr, N. (1957) *Atomic Physics and Human Knowledge.* New York: Wiley and Sons 1958. (Original in Danish, 1957).

Bohr, N., Kramers, H. A. and Slater, J. C. (1924) The quantum theory of radiation. *The Philosophical Magazine,* 47, 785-822.

Bohr, N. (1972) *Collected Works.* Ed. by L. Rosenfeld, and J. R. Nielsen. Amsterdam: North Holland/New York: Elsevier. vol. 1, 1972.

Boltzmann, L. (1896/1898) *Vorlessungen über Gastheorie I. Theil.* Leipzig: Barth 1896. *Ibid., II. Theil,* 1898. Engl. transl. by S. G. Brush, *Lectures on Gas Theory.* Berkeley, 1964.

Born, M. (1926) Quanten Mechanik der Stössvorgänge, *Zeitschrift für Physik* 38, 1926, 803-827; Also in Born (1963) vol. 2, 233-258. Engl. transl., Quantum mechanics of collision processes. In G. Ludwig (Ed.), *Wave Mechanics.* London: Pergamon Press, 1968.

Born, M. (1927) Quantenmechanik und Statistik, *Naturwissenschaftlich* 15, 238-242.

Born, M. (1963) *Ausgewählte Abhandlungen.* Göttingen: Vandenhoeck & Ruprecht, 2 vols.

Born, M. and Heisenberg, W. (1928) La mécanique des quanta. In *Electrons et Photons,* 143-183.

Bothe, W. and Geiger, H. (1924) Ein Weg zu experimentellen Nachprüfung der Theorie von Bohr, Kramers und Slater, *Zeitschrift für Physik* 26, 1924, 44. Engl. transl.: Experimental test of the theory of Bohr, Kramers and Slater. In Lindsay (1979), 230-231.

Bothe, W. and Geiger, H. (1925) Über das Wesen des Comptoneffekts; eine experimentelles Beitrag zur Theorie des Strahlung, *Naturwissenschaft* 13, 440-; *Zeitschrift für Physik* 32, 639-663.

Bunge, M. (1985) *Treatise on basic philosophy,* Vol. 7: *Epistemology and methodology,* III: *Philosophy of science and technology.* Part I: *Formal and physical sciences.* Dordrecht: Reidel.

Darrigol, O. (1988) Statistics and combinatorics in early quantum theory. *Historical Studies in the Physical Sciences,* 19, 17-80.

Darrigol, O. (1991) Statistics and combinatorics in early quantum theory, II: Early symptoma of indistiguishability and holism. *Historical Studies in the Physical Sciences* 21, 237-298.

De Broglie, L. (1924) *Recherches sur la théorie des quanta,* Thèse, Paris, 1924; *Annales de Physique,* 10th series, 3, 1925, 22-128; reprinted Paris: Masson (1963).

Dirac, P. A. M. (1930) *The Principles of Quantum Mechanics*. Oxford: Clarendon Press. 4th ed., 1958.

Ehrenfest, P. and T. (1911) Begriffliche Grundlagen der Statistischen auffassung in der Mechanik. In *Encyclopadie der mathematischen Wissenschaften*. Leipzig: Teubner, vol. 6, *Mechanik*, part 6, 1-90.

Ehrenfest, P. (1912) Zur Frage der Entbehrlichkeit der Lichtäthers, *Physikalische Zeitschrift* 13, 317-319.

Ehrenfest, P (1959) *Collected Scientific Papers*, edited by M. Klein. Amsterdam: North Holland.

Einstein, A. (1903) Eine Theorie der Grundlagen der Thermodynamik, *Annalen der Physik*, ser. 4, 11, 170-187. (CP, 2, 77-94).

Einstein, A. (1905a) Ueber die von der molekular kinetischen Theorie der Wärme geforderte Bewegung von in ruhenden Flüssigkeiten suspendierten Teilchen, *Annalen der Physik*, ser. 4, 17, 549-560. (CP, 2, 224-235).

Einstein, A. (1905b) Ueber einen die Erzeugung und Verwandlung des Lichtes betreffenden heuristischen Gesischtspunkt, *Annalen der Physik*, ser. 4, 17, 132-148. (CP, 2, 150-166).

Einstein, A. (1906) Zur Theorie der Lichterzeugung und Lichtabsorption, *Annalen der Physik*, ser. 4, 20,199-206. (CP, 2, 350-357).

Einstein, A. (1907) Die Planck'sche Theorie der Strahlung und die Theorie der specifischen Wärme, *Annalen der Physik*, ser. 4, 22, 180-190. (CP, 2, p. 379-391).

Einstein, A. (1909a) Zum gegenwärtigen stand des Strahlungsprobleme, *Physikalische Zeitschrift* 10, 185-193. (CP, 2, 542-553).

Einstein, A. (1909b) Ueber die Entwicklung unserer Anschauungenüber das Wesen und die Konstitution der Strahlung, *Deutsche Physikalische Gessellschaft, Verhandlungen* 7, 1909, 482-500 (Salzbourg Conference, 21 nov. 1909); also, *Physikalische Zeitschrift* 10, 1909, 817-825. (CP, 2, 564-583).

Einstein, A (1912g) Etat actuel du problème des chaleurs spécifiques. In M. de Broglie and P. Langevin (Eds.), *La théorie du rayonnement et les quanta. Communications et discussions de la réunion tenue à Bruxelles du 30 octobre au 3 novembre 1911, sous les auspices de M.E. Solvay*. Paris: Gauthier-Villars, 407-450. (French transl. by P. Langevin. German original, CP, 3, p. 521-548.

Einstein, A. (1916-1917) Zur Quantentheorie der Strahlung, *Physikalische Gesellschaft Mitteilungen* (Zürich), 47-62; also *Physikalische Zeitschrift* 17, 121-128. Engl. transl. in Waerden, 1967.

Einstein, A. (1924-1925) Quantentheorie des einatomigen idealen Gases I and II, *Preussische Akademie Wissenschaften, Phys. Math. Klasse, Sitzungsberichte* 22, 261-267; 23, 3-14.

Einstein, A. (1925) Quantentheorie des idealen Gases, *Preussische Akademie Wissenschaften, Phys. Math. Klasse, Sitzungsberichte*, 23, 18-25.

Einstein, A. (1926) Observações sobre a situação actual da theoria da luz, *Revista da Academia Brasileira de Sciencias*, 1, 1-3. German original: Bemerkungen zu der gegenwärtigen Lage der Theorie des Lichtes. In A. T. Tomlasquin and I. C. Moreira (Eds.), *Ciência hoje* (Rio de Janeiro), vol. 21, 1997, 124, 25-27.

Einstein, A. (1987-1998) *The Collected Papers of Albert Einstein*, J. Stachel, D. C. Cassidy, R. Schulmann, M. Klein et al. (Eds.), Princeton: Princeton University Press, 8 publ. vols. (referred to as CP)

Einstein, A. and Born, M. (1969) *Briefwechsel 1916-1955*, München: Nymphenburger Verlagshandlung.

Einstein, A. and Langevin, P. (1972) Correspondance. In L. Langevin, Paul Langevin et Albert Einstein d'après une correspondance et des documents inédits, *La Pensée* 161, 3-40.

Electrons, Photons (1928). *Electrons et photons. Rapports et discussions du cinquième Conseil de physique tenu à Bruxelles du 24 au 29 octobre 1927 sous les auspices de l'Institut international de physique Solvay*. Paris: Gauthier-Villars.

Gibbs, W. (1902) *Elementary Principles in Statistical Mechanics*, New Haven: Yale University Press.

Gould, C. (Ed.) (forthcoming) Constructivism and Practice: Towards a Social and Historical Epistemology. A volume dedicated to Marx Wartofsky's memory. Lanham, MD: Rowman & Littlefield.

Jammer, M. (1966) *The Conceptual Development of Quantum Mechanics*. New York: McGraw-Hill.

Jammer, M. (1974) *The Philosophy of Quantum Mechanics: The Interpretations of Quantum Mechanics in Historical Perspective*. New York: Wiley and Sons.

Kastler, A. (1983) On the historical development of the indistinguishability concept for microparticles. In A. van der Merwe (1983), 607-623.

Klein, M. J. (1962) Max Planck and the beginning of the quantum theory. *Archive for the History of Exact Science* 1, 459-479.

Klein, M. J. (1965) Einstein, specific heats and the early quantum theory. *Science*, 148, 173-180.

Kolmogorov, A. N. (1933) *Grundbegriffe der Wahrscheinlichkeitsrechnung*, Berlin: Springer Verlag, English transl. by N. Morrison, *Foundation of the Theory of Probability*, New York: Chelsea, 1954.

Korner, S. (Ed.) (1957) *Observation and Interpretation in the Philosophy of Physics. With Special Reference to Quantum Mechanics*, New York: Dover.

Kuhn, T. (1978) *Black-body Theory and the Quantum Discontinuity, 1894-1912*. New York: Clarendon Press.

Langevin, P. (1936) *La Notion de Corpuscule et d'Atome*, Paris: Hermann.

Langevin, P. and De Broglie, M. (Eds.) (1912) *La théorie du rayonnement et les quanta. Communications et discussions de la réunion tenue à Bruxelles du 30 octobre au 3 novembre 1911, sous les auspices de M.E. Solvay*, Paris: Gauthier-Villars.

Laplace, P. S. (1814) *Essai philosophique sur les probabilités*, augm, ed. (1825), in *Oeuvres,* vol 7.

Laplace, P. S. (1878-1912) *Oeuvres,* 14 vols., Paris: Gauthier-Villars.

Lindsay, R. (Ed.) (1979) *Early Concepts of Energy in Atomic Physics.* Stroudsbury, Penn: Dowden, Hutchinson and Ross.

Lopes, J. L. and Paty, M. (Eds.) (1977) *Quantum Mechanics, a Half Century Later.* Dordrecht: Reidel.

Mugur-Schächter, M. (1977) the quantum mechanical one-system formalism, joint probabilities and locality. In J. L. Lopes and M. Paty, *Quantum Mechanics, a Half Century Later.* 107-166.

Natanson, L. (1911) On the statistical theory of radiation, *Bulletin de l'Académie des sciences de Cracovie* (A), 134-138. Ueber die statistische Theorie der Strahlung, *Physikalische Zeitschrift* 12, 1911, 659-666.

Paty, M. (1988) *La matière dérobée. L'appropriation critique de l'objet de la physique contemporaine*, Paris: Archives Contemporaines.

Paty, M. (1990) Reality and Probability in Mario Bunge's *Treatise.* In G. Dorn and P. Weingartner (Eds.), *Studies on Mario Bunge's Treatise.* Amsterdam-Atlanta: Rodopi, 301-322.

Paty, M. (1993a) Sur les variables cachées de la mécanique quantique: Albert Einstein, David Bohm et Louis de Broglie, *La Pensée*, 292, 93-116.

Paty, M. (1993b) Formalisme et interprétation physique chez Schrödinger. In M. Bitbol and O. Darrigol (Eds.), *Erwin Schrödinger. Philosophy and the Birth of Quantum Mechanics. Philosophie et naissance de la mécanique quantique*, Paris: Editions Frontières, 161-190.

Paty, M. (1995) The nature of Einstein's objections to the Copenhagen interpretation of quantum mechanics, *Foundations of Physics,* 25, 183-204.

Paty, M. (1999) Are quantum systems physical objects with physical properties? *European Journal of Physics,* 20, 373-388.

Paty, M. (2000a) Interprétations et significations en physique quantique, *Revue Internationale de philosophie,* 212, 199-242..

Paty, M (2000b) The quantum and the classical domains as provisional parallel coexistents, *Synthese*, 125, 179-200. In honor of Newton da Costa, on the occasion of his seventieth birthday, ed. by French, S., D. Krause, F. Doria..

Paty, M. (forthcoming a) *Einstein, les quanta et le réel.*

Paty, M. (forthcoming b) The concept of quantum state: New views on old phenomena. In R. S. Cohen, D. Howard, J. Renn, S. Sarkar and A. Shimony, (Eds.), *John Stachel Festschrift*, Boston Studies in the Philosophy and History of Science. Dordrecht: Kluwer.

Paty, M. (forthcoming c) The idea of quantity at the origin of the legitimacy of mathematization in physics. In C. Gould (Ed.), *The Philosophy of Marx Wartofsky*, New York.

Planck, M. (1900) Zur Theorie des Gesetzes der Energieverteilung im Normalspektrum, *Verhandlungen der deutschen physikalischen Gessellschaft* 2, 237-245. Also, with Engl. transl., On the theory of the energy distribution law of the normal spectrum, In M. Planck (1972), 38-45.

Planck, M. (1972) *Planck's Original Papers in Quantum Physics*, German and English ed., annotated by H. Kangro, transl. by D. ter Haar and S. G. Brush, London: Taylor and Francis.

Popper, K. (1957) The propensity interpretation of the calculus of probability and the quantum theory. In S. Körner (Ed.), *Observation and Interpretation*, London: Butterworths.

Popper, K. R. (1982) *Quantum Theory and the Schism in Physics, from the Postscript to the Logic of Scientific Discovery*, vol. 3. London: Hutchinson.

Popper, K. R. (1990) *A World of Propensities*. Bristol: Thoemmes Press.

Reichenbach, H. (1944) *Philosophical Foundations of Quantum Mechanics*. Berkeley: University of California Press.

Reichenbach, H. (1978) *Selected Writings*, M. Reichenbach and R. S. Cohen (Eds.), 2 vols. Dordrecht: Reidel.

Rosenfeld, L. (1974) Statistical causality in atomic theory. In *Elkana 1974*, 469-480.

Rutherford, E. (1952/1965) *The Collected Papers of Lord Rutherford of Nelson*, publ. under the scientific direction of James Chadwick, 3 vols. London: Allen and Unwin.

Schrödinger, E. (1926) *Abhandlungen zur Wellenmechanik*, Leipzig: Barth

Shushurin, S. P. (1977) Essay on the development of the statistical theory of the calculus of probability. In Lopes and Paty, 89-106.

Suppes, P. (1961) Probability concepts in quantum mechanics, *Philosophy of science* 28, 378-389.

Suppes, P. (1963) The role of probability in quantum mechanics. In B. Baumrin, (Ed.), *Philosophy of Science, The Delaware Seminar*. New York: Wiley.

Suppes, P (1970) *A Probablistic Theory of Causality*. Amsterdam: North Holland.

Toraldo di Francia, G. (Ed.) (1979) *Problems in the Foundations in Physics. Rendiconti della Scuola Internazionale di Fisica "Enrico Fermi", 72 Corso, Varenna, 1977*, Bologna: Soc. Ital. di Fisica/Amsterdam: North Holland.

Van der Merwe, A. (Ed.) (1983) *Old and New Questions in Physics, Cosmology, Philosophy and Theoretical Biology*. New York: Plenum Press.

Van der Waerden, B. L. (Ed.) (1967) *Sources of Quantum Mechanics*. Amsterdam: North Holland.

Von Neumann, J. (1932) *Mathematische Grundlagen der Quantenmechanik*, Berlin: Springer.

15

Stochastic Dynamical Reduction and Causality

GianCarlo Ghirardi

University of Trieste

1 Introduction

The very title of this conference seems, from the perspective of my scientific interests, to address precisely the core of the foundational problems of quantum mechanics. Such a theory, one of the pillars of modern science, contains irreducible *stochastic elements* which are specifically related to the measurement process (more generally to the objectification problem). Moreover, it is precisely with reference to such a process which, when one takes further into account the unavoidable *non-local character* of the theory, the problem of the *causal relations between events* emerges as a central one.

An adequate analysis of such a problem requires a relativistic approach. Accordingly, the subject of my analysis will be the problem of *statevector reduction in relativistic quantum mechanics*. I will also discuss the related problems of *property attribution* to individual physical systems and of *counterfactual arguments* within a relativistic framework, by making specific reference to the *Dynamical Reduction Program*.

Stochastic Causality.
Maria Carla Galavotti, Patrick Suppes and Domenico Costantini (eds.).

2   The Reduction Process and Its Difficulties

As is well known, the reduction (or macro-objectification) process represents a crucial point of nonrelativistic quantum formalism. Further difficulties arise in connection with relativistic considerations. Let us briefly review them.

2.1. A Crucial Point: The Instantaneity of State Vector Reduction

According to the *textbook* formulation of the rules of the game (nonrelativistic quantum mechanics) one adopts the *Reduction Postulate:*

A measurement process causes (in general) the instantaneous jumping of the statevector before the measurement into the linear manifold associated to the eigenvalue of the measured observable corresponding to the outcome of the measurement.

As a simple example we can consider (oversimplifying the problem) the linear superposition of the improper position eigenstates $|x_1>$ and $|x_2>$ of a particle subjected to a position measurement. Denoting as $|\Psi_{before}>$ and $|\Psi_{after}>$ the statevector before and after the measurement, one has:

$$|\Psi_{before}> = \frac{1}{\sqrt{2}} \left[|x_1> + |x_2>\right], \quad |\Psi_{after}>, \text{ either } |x_1> \text{ or } |x_2>. (1)$$

The question we want to analyze is then: *How can the above description be generalized within a relativistic context?* We stress that this is a highly nontrivial problem for the very reason that even the postulate of wave packet reduction within the nonrelativistic version of the theory contradicts the idea that the measurement itself be a quantum process. This is the main reason for which, in the final part of the paper, I will resort to the consideration of the only (to my knowledge) consistent theoretical schemes accounting for reductions, i.e., the so called GRW and CSL theories.

2.2. An Elementary Example

We begin by making clear that we will often disregard (when this does not invalidate our considerations) the spreading of the wave function and that in this first part of the analysis the wavefunction $\Psi(x,t)$ is considered as a function on the space-time continuum.

Our first example is the one of a free particle which is described (in a given reference frame) by a wavefunction having, at any time $t < 0$, an appreciable extension, and which is subjected, at $t = 0$, to a position measurement. We suppose that the particle is found within a region Δ around the origin. One immediately meets some problems. Even if we describe the local interaction between the detector and the particle in a covariant way, the reduction process is not covariant. In fact, in a moving frame the space-like surface $t = 0$ becomes a space-like surface σ which is not an equal-time hypersurface.

2.3. A Proposal for a Covariant Reduction Process

Hellwig and Kraus (1970) have put forward a suggestion to overcome the above difficulty. The idea is very simple: *In a local measurement, reduction takes place along the backward light cone from the region of the measurement.*

We mention that there have also been proposals of a manifestly noncovariant description, like the one by Cohen and Hiley (1995), which assume that the reduction takes place instantaneously in a preferred reference frame. We do not share such a position and we consider it useful to discuss in all details the proposal of Hellwig and Kraus (1970). To this purpose we have to make a short digression, i.e., we have to distinguish between what have been denoted as *Local* and *Nonlocal* Measurements.

2.4. *Local* Versus *Nonlocal* Measurements

The first question we want to face is: w*hy is it important to distinguish between such measurements?* To clarify this point we start by remarking that, as a consequence of the indeterminacy relations, certain types of measurement processes imply an instantaneous and arbitrary increase of the spread of the wavefunction.

Let us consider, e.g., a wavefunction having compact support at $t = 0$. In a relativistic framework it must remain concentrated within the forward light cone from its support. At time t we perform a momentum measurement with such an accuracy that the relation $\Delta x \Delta p \geq \hbar/2$ implies an instantaneous and extremely relevant increase of Δx. This shows that the momentum measurement induces an appreciable probability of finding, at time $t + \varepsilon$, the particle at a point P in which it could not be found if no momentum measurement would have been performed.

The conclusion should be obvious: *The (nonlocal) momentum measurement allows (in principle) faster than light signaling.* Actually, this embarrassing situation has already been identified by Landau and Peierls (1931) and has led them to conclude that: *within a relativistic framework the possibility of nonlocal measurements has to be excluded* [1].

---

[1] Incidentally, it is useful to note that Bohr and Rosenfeld (1933) with their analysis of the possibility of measuring operators of the electromagnetic field respecting the indeterminacy

2.5. The Implications of Nonlocal Measurement Processes

A more convincing example of the disrupting effects of an unavoidably *nonlocal* measurement is the one of the measurement of the square of the angular momentum of a composite system whose constituents are far-apart (let us say at x_1 and x_2, respectively). The example involves two particles of spin (or isospin) 1/2 in the $|1,1>$ state, a device which allows, at the experimenter's will, to induce a spin-flip of one of the two particles and an apparatus measuring nonlocally the square of the total spin.

We recall that the spin states for such a system have, in the uncoupled and the coupled representations, respectively, the form:

$$|\uparrow>_{x1}|\uparrow>_{x2}, \quad |\downarrow>_{x1}|\downarrow>_{x2}, \quad |\uparrow>_{x1}|\downarrow>_{x2}, \quad |\downarrow>_{x1}|\uparrow>_{x2}, \tag{2}$$

and

$$|1,1>=|\uparrow>_{x1}|\uparrow>_{x2}, \quad J^2=1, J_z=1$$

$$|1,-1>=|\downarrow>_{x1}|\downarrow>_{x2}, J^2=1, J_z=-1$$

$$|1,0>=\frac{1}{\sqrt{2}}\left[|\uparrow> x1\,|\downarrow> x2+|\downarrow\, x1\,|\uparrow> x2\right], \quad J^2=1, J_z=0 \tag{3}$$

$$|0,0>=\frac{1}{\sqrt{2}}\left[|\uparrow> x1\,|\downarrow> x2-|\downarrow\, x1\,|\uparrow> x2\right], \quad J^2=0, J_z=0$$

One can then argue as follows:

i. A measurement of J^2 is performed at time t.

ii. At time $t+\varepsilon$ a measurement of $\sigma_z^{(x2)}$ is performed at x_2.

iii. At time $t-\varepsilon$ a spin-flip of particle x_1 can be induced or not by a local device, at the experimenter's free-will.

Then:

i. If no spin-flip takes place, the measurement at t gives the outcome 1 with certainty and does not affect the state of the two particles. Consequently, the measurement at $t+\varepsilon$ gives, with certainty, the outcome $+1$.

---

relations implied by the commutation relations between field operators, had already, in some sense, "disproved" the assumption of Landau and Peierls.

ii.   If the spin-flip occurs, then the state before t is:

$$|\downarrow>_{x1}|\uparrow>_{x2} = \frac{1}{\sqrt{2}}[|1,0> - |0,0>],$$ (4)

so that both outcomes $J^2 = 0,1$ can occur with equal probabilities. No matter which one is obtained, the probability of getting the result $+1$ at $t + \varepsilon$ is $1/2$.

Concluding: *The nonlocal measurement of J^2 at time t, allows one to change appreciably, by performing or not an action at the point $(x_1, t - \varepsilon)$, the probability of the outcome at $(x_2, t + \varepsilon)$. Since the separation between such points is space-like, faster than light signaling becomes possible.*

The relevance of the previous remarks for the proposal of Hellwig and Kraus (1970) is evident. However it has to be stressed that these authors have proved that if (in accordance with the position of Luders and Peierls (1931)) one forbids nonlocal measurements, then their suggestion does not lead to any contradiction.

Obviously a crucial question remains: is it true that nonlocal measurements are forbidden? Aharonov and Albert (1980, 1981) have analyzed this point in great detail, arriving at the important conclusion that one can measure *locally* a *nonlocal* observable. Their analysis implies that the proposal by Hellwig and Kraus is not tenable.

3   Performing Nonlocal Measurements in a Local Way

The explicit proof that one can perform locally measurements of nonlocal observables has far reaching consequences for the problem of working out a relativistic theory of wave packet reduction. Let us briefly recall the conclusion drawn by Aharonov and Albert and stress their implications.

3.1. A Concise Summary of the Work by Aharonov and Albert

Aharonov and Albert (1980,1981), and the same authors together with Vaidman (1986), have performed an interesting analysis of the possibility of performing nonlocal measurement by resorting to local system-apparatus interactions. Their proposal is very smart and requires a subtle utilization of quantum entanglement: a (possibly microscopic) part of the apparatus is represented by a composite system whose components are entangled.

For lack of space we simply list the most important implications of their work, and we discuss its most relevant implications in the following subsections. They have proved:

a)   That genuinely nonlocal measurements can be performed by means of local interactions. This very fact represents a deadlock for proposals of the Hellwig and Kraus type.

2. That a consistent way to deal with a relativistic reduction scheme requires adopting a completely new point of view about relativistic wavefunctions. They are no more functions on the space-time continuum but functions on the set of space-like surfaces, so that we must write

$$\Psi = \Psi(\sigma, x) \tag{5}$$

where x runs over the points of the space-like hypersurface σ.

We stress that this idea is strictly related to the old and brilliant suggestion of describing the quantum evolution according to the Tomonaga-Schwinger formalism. However, some relevant differences have to be outlined since the above authors were not contemplating actual reduction processes to take place (more about this in what follows).

3. That, with the new perspective, the proposal of Aharonov and Albert, which could summarized by saying that in a local measurement reduction takes place instantaneously in all inertial frames, i.e., along all space-like surfaces going through the space-time point where the local interaction occurs, leads to a consistent description of reduction processes even within a relativistic context in which measurements of nonlocal observables are possible.

3.2 The Implications of the Aharonov and Albert Analysis for the Statevector

Aharonov and Albert (1980, 1981), having shown that one can perform nonlocal measurements by means of local interactions have suggested a new covariant reduction mechanism. Their proposal is quite simple and natural but has far reaching consequences.

Let us consider a space-time point P where a local measurement occurs and the set of all space-like planes through P and let us notice that each such plane is a constant time space-like surface for an appropriate observer. The basic assumption is then that, *for the considered observer, reduction takes place just along that plane (more generally they consider arbitrary space-like surfaces such that the normal to them at P coincides with the time axis of the associated observer).*

We note that the rule is manifestly covariant and quite natural, since the main interest of the theory consists in its allowing each observer to make predictions for future (for him) events. *However, it has to be stressed that such a game is possible only at the price of a radical change of perspective concerning the statevector.* This is easily seen by considering a point P where a local reduction occurs, and two hyperplanes σ_1 and σ_2 intersecting at a point Q such that P belongs to the region $\sigma_1 \hat{Q} \sigma_2$, with obvious meaning of the symbols. Both hyperplanes are equal-time planes for appropriate observers. For the observer for which σ_1 is such a hyperplane the local measurement has not occurred (it will occur on a subsequent hyperplane), consequently, the statevector is unreduced. For the other, reduction has already taken place and the statevector is reduced.

Summarizing, the conceptually more relevant consequence of the Aharonov and Albert proposal is that *the wavefunction $\Psi(x,t)$ cannot be thought any more as a function on the space-time continuum, but as a function on the set of space-like surfaces:*

$$\Psi(\sigma,x), \quad x \in \sigma. \tag{6}$$

We note that this implies that, for the same objective point $\tilde{Q}$, and for 2 surfaces σ_1 and σ_2 through it, one must accept that:

$$\Psi(\sigma_1,\tilde{Q}) \neq \Psi(\sigma_2,\tilde{Q}). \tag{7}$$

Two remarks are appropriate:

1. Also within the Tomonaga-Schwinger approach to relativistic quantum field theory one associates different statevectors to different surfaces even when they have some common portion. However, within the considered theoretical framework, the expectation value of any (local) observable having its support confined to the common support of the two considered hypersurfaces has the same value, independently of which of the two statevectors one takes into account. In particular, therefore, the local (improper) position observable would have equal probabilities of giving the outcome $\tilde{Q}$ independently of which surface one takes into account. This is a direct consequence of the fact that the Tomonaga-Schwinger theory describes the linear and deterministic evolution of quantum systems and does not account for nonlinear and stochastic processes like the reduction of the wave packet. In particular, eq. (7) does not fit within a Tomonaga-Schwinger scheme. The scheme itself must be radically changed to allow for reductions.

2. Aharonov and Albert do not introduce any explicit consistent dynamical mechanism leading to what they assume to occur in a measurement process. To achieve this one has to identify a new universal dynamical mechanism. This is precisely the key point of Dynamical Reduction Models.

4   The Dynamical Reduction Program

As already stated the measurement problem presents serious conceptual difficulties at the nonrelativistic level. A proposal to circumvent them is the so called GRW theory (Ghirardi, Rimini and Weber 1986) and its continuous generalization, CSL (Pearle 1989; Ghirardi, Pearle and Rimini 1990). Let us briefly review these theories.

4.1. The GRW Theory

As pointed out by J. S. Bell (1987), this approach corresponds to accepting that *Schrödinger's equation is not always right* and introducing stochastic and nonlinear modification to it in order to allow a unified treatment of all natural processes.

The GRW model is based on the assumption that, besides the standard evolution, physical systems are subjected to spontaneous localizations occurring at random times and affecting their elementary constituents. Such processes are formally described in the following way.

We consider a system of N particles. We suppose that when the i-th particle of the system suffers a localization the wave function changes according to

$$\Psi(r_1,...,r_N) \to \Psi_x(r_1,...,r_N) = \Phi_x(r_1,...,r_N)/\|\Phi_x\|$$

$$\Phi_x(r_1,...,r_N) = (\alpha/\pi)^{3/4} e^{-\frac{\alpha}{2}(r_i-x)^2} \Psi(r_1,...,r_N). \tag{8}$$

The probability density of the process occurring at point x is given by $\|\Phi_x\|^2$. For what concerns the temporal features of the processes we assume that the localizations of the various particles occur independently and at randomly distributed times with a mean frequency λ_m which depends on their mass. We choose $\lambda_m = (m/m_0)\lambda$, where m is the mass of the particle, m_0 is the nucleon mass and λ is of the order of $10^{-16} \sec^{-1}$. The localization parameter $1/\sqrt{\alpha}$ is assumed to take the value $10^{-5} cm$.

The most appealing feature of the model derives from its *trigger mechanism*: in the case of a macroscopic system the localization processes are amplified with the number of their constituents. Accordingly, superpositions of states corresponding to different locations of a macroscopic object cannot persist for *more than a split second* (Bell 1987).

4.2. The CSL Theory

The model is based on a linear stochastic evolution equation for the state-vector (Pearle 1989; Ghirardi, Pearle and Rimini 1990). The evolution does not preserve the norm but only the average value of the squared norm. The equation is:

$$\frac{d \,|\, \Psi_w(t) >}{dt} = [-\frac{i}{h} H + \sum_i A_i w_i(t) - \gamma \sum_i A_i^2] \,|\, \Psi_w(t) > . \qquad (9)$$

The quantities $w_i(t)$ are c-number Gaussian stochastic processes satisfying

$$<< w_i(t) >> = 0, << w_i(t) w_j(t) >> = \gamma \delta_{ij} \delta(t - t'), \qquad (10)$$

while the quantities A_i are commuting self-adjoint operators. Let us assume, for the moment, that these operators have a purely discrete spectrum and let us denote by P_σ the projection operators on their common eigenmanifolds.

The physical meaning of the model is made precise by the following prescription: if a homogeneous ensemble (pure case) at the initial time $t=0$ is associated to the statevector $|\Psi(0)>$, then the ensemble at time t is the union of homogeneous ensembles associated with the normalized vectors $|\Psi_w(t)>/|||\Psi_w(t)>||$, where $|\Psi_w(t)>$ is the solution of the evolution equation with the assigned initial conditions and for the specific stochastic process w which has occurred in the interval $(0,t)$. The probability density for such a subensemble is the *cooked one*, i.e., it is given by:

$$P_{Cook}[w] = P_{Raw}[w]||\quad \quad |\Psi_W(t)>||^2, \qquad (11)$$

where we have denoted by $P_{Raw}[w]$ the *raw* probability density associated to the Gaussian process w in the interval $(0, t)$, i.e.,:

$$P_{Raw}[w] = \frac{1}{N} e^{-\frac{1}{2\gamma} \sum_i \int_0^t d\tau w_i^2(\tau)}, \qquad (12)$$

N being a normalization factor.

One can prove that the map from the initial to the final ensemble obeys the forward time translation semigroup composition law. It is also easy to prove that the evolution, at the ensemble level, is governed by the dynamical equation for the statistical operator

$$\frac{d\rho(t)}{dt} = \frac{i}{\hbar}[\rho(t), H] + \gamma \sum_i A_i \rho(t) A_i - \frac{\gamma}{2}\{\sum_i A_i^2, \rho(t)\}, \tag{13}$$

from which one immediately sees that, if one disregards the Hamiltonian evolution, the off-diagonal elements $P_\sigma \rho(t) P_\tau, (\sigma \neq \tau)$ of the statistical operator are exponentially damped.

The final step consists in identifying the discrete index i and the operators A_i in the above equations with the continuous index $\mathbf{r}$ and the average mass density operator:

$$M(\mathbf{r}) = \sum_k m^{(k)} N^{(k)}(\mathbf{r}),$$

$$N^{(k)}(\mathbf{r}) = \left[\frac{\alpha}{2\pi}\right]^{\frac{3}{2}} \sum_s \int d\mathbf{q}\, e^{-\frac{\alpha}{2}(\mathbf{q}-\mathbf{r})^2} a_{(k)}^\dagger(\mathbf{q}, s) a_{(k)}(\mathbf{q}, s), \tag{14}$$

where the operators $a_{(k)}^\dagger(\mathbf{q},s)$ and $a_{(k)}(\mathbf{q},s)$ are the creation and annihilation operators of a particle of type k (k = proton, neutron, electron,...) at point $\mathbf{q}$ with spin component s. With these choices it is easy to show that the theory implies that any macroscopic system is always extremely well localized in space and that when a superposition of microstates triggers a macroscopic object (typically a measuring apparatus) the different final locations of its pointer reproduce the quantum probabilities of the measurement outcomes.

4.3. Relativistic CSL Theory

To get a relativistic generalization of the CSL theory one adopts the quantum field theoretical point of view (Pearle 1990; Ghirardi, Grassi and Pearle 1990). The stochasticity of the evolution is accounted by a stochastic interaction term in the Lagrangian density. As an example of the procedure we consider the Lagrangian density:

$$L(x) = L_0(x) + L_I(x)V(x), \tag{15}$$

where

$L_0(x)$ and $L_I(x)$, are scalar functions of the fields, $L_I(x)$ does not contain derivative couplings and $V(x)$ is a c-number stochastic process which is a scalar for Poincaré transformations. We choose for $V(x)$ a Gaussian noise with mean zero. To have a relativistic stochastically invariant theory, its covariance must be an invariant function:

$$<< V(x)V(x') >> = A(x-x'), A(\Lambda^{-1}x) = A(x) . \tag{16}$$

In the literature, the choice:

$$A(x-x') = \lambda\delta(x-x') \tag{17}$$

has been made. It gives rise to specific problems related to the appearance of untractable divergences in the formalism. Here we will disregard these problems and we will concentrate our attention on the fact that, formally, the scheme, as it has been shown in Ghirardi, Grassi and Pearle (1990), is perfectly consistent and satisfactory from the point of view of its relativistic aspects.

4.4. The Tomonaga-Schwinger Version of CSL

To exhibit explicitly the relations with the Tomonaga-Schwinger formalism we present now the version of relativistic CSL which has been proposed in Pearle (1990) and Ghirardi, Grassi and Pearle (1990). The formalism can be summarized in the following terms:

1. The fields are solutions of the Heisemberg equations associated to $L_0(x)$.

2. The statevector evolution is governed by the equation

$$\frac{\delta | \Psi_V(\sigma) >}{\delta\sigma(x)} = \left[L_I(x)V(x) - L_{I(x)}^2 \right] | \Psi_V(\sigma) > . \tag{18}$$

Note the skew-hermitian character of the coupling to the stochastic c-number field. The corresponding equation for the statistical operator is:

$$\frac{\delta\rho(\sigma)}{\delta\sigma(x)} = \lambda L_I(x)\rho(\sigma)L_I(x) - \frac{\lambda}{2}\left\{ L_i^2(x), \rho(\sigma) \right\}, \tag{19}$$

from which one can derive the reduction properties of the theory.

3. The following choices are made for the Lagrangian densities $L_0(x)$ and $L_I(x)$:

$$L_0(x) = \frac{1}{2}\left[\partial_\mu\Phi(x)\partial^\mu\Phi(x) - m^2\Phi^2(x) \right] + \overline{\Psi}(x)\left[i\gamma^\mu\partial_\mu - M \right]\Psi(x)$$
$$+ \eta\overline{\Psi}(x)\Psi(x)\Phi(x), \tag{20a}$$

$$L_I(x) = \Phi(x). \tag{20b}$$

With these premises one easily proves that the dynamics leads to a localization of the fermions associated to the field $\Psi(x)$. In the nonrelativistic limit (ignoring the problem of the divergences) the framework exhibits features which are quite similar to those of the CSL model. The reasons for this fact are easily understood: differently located fermions carry with them different virtual mesonic clouds and the dynamics does not tolerate the superpositions of such clouds.

Two concluding remarks seem appropriate:

1. Due to the reducing features of the theory, the states associated to different hypersurfaces can lead to different expectation values at the same point, just as in the case of Aharonov and Albert.

2. The theory does not exhibit time reversal invariance (only the forward time translation part of the Poincaré semigroup is represented). Accordingly, the initial condition must be fixed on a precise, objective, spacelike hypersurface. The passage to other observers requires the adoption of the passive point of view.

5 Events and Properties in a Quantum Context

Before proceeding it is important to focus some crucial features concerning physical events and the possibility of attributing objective properties to individual physical systems in a genuinely relativistic and nonlocal quantum context. To this purpose let us consider a state:

$$|\Psi, t> = \alpha\,|\omega_1> + \beta\,|\omega_2> \tag{21}$$

$|\omega_i>, (i = 1, 2)$ being the eigenvectors of an observable Ω of the system belonging to different eigenvalues. Such a state shows that the very concept of a quantum object enters into an irredeemable conflict with the one deriving from our spatio-temporal and causal intuition at the macroscopic level: the quantum level of reality is characterized by an *objective indefiniteness* of most properties. There is an underlying structure (which we successfully describe by our formalism) possessing *potential properties* which, in general, are neither actual nor referring to individuals (due to the indistinguishability of identical microcostituents and to the entanglement of composite quantum systems).

Abner Shimony (1986) has been particularly lucid in stressing this unavoidable conceptual tension between the features of physical processes imposed by their quantum nature and our familiar way of describing them within space-time. He points out that a new modality of reality emerges at the quantum level lying in a certain sense between the pure logical possibility and the full actuality of a spatio-temporal event. With respect to the causal spatio-temporal description such a reality possesses the ontological status of a potentiality. However, since here we are interested in the relativistic aspects of the problem, it seems extremely relevant to call attention on an important remark by Shimony himself.

Shimony claims that the quantum conception of nature *may leave relativistic space-time structure intact, but only change our conception of an event in space-time.* And he proposes a *definition* of an *event* fitting perfectly the quantum context (both relativistic and nonrelativistic): *the definiteness or indefiniteness of a physical variable constitutes an event.*

Correspondingly one has to be careful in characterizing the properties which can be considered as objectively possessed by an individual physical system. This is obtained by resorting to the following criterion:

An individual physical system has at a given time t an objective physical property referring to the observable Ω *iff the statevector* $|\Psi, t >$ *of the system at the considered time is an eigenvector of* Ω *corresponding to a certain eigenvalue* ω_k.

We note that in such a case the event is: Ω is definite and has the value ω_k, while, in a situation like the one of eq. (21) the event is: Ω is indefinite.

6 Events and Criteria for Property Attribution in a Relativistic Context

Keeping in mind what we have learned from the previous analysis we can now face the central problem of identifying the basic features and of elaborating the appropriate conceptual framework for dealing with a relativistic theory inducing reduction.

6.1. A Relativistic Simplified Dynamical Reduction Model

We could perform our analysis by making reference to the relativistic CSL model presented in Subsection. 4.4. However, for the sake of simplicity, we take into consideration an extremely oversimplified model. Let us describe it.

1. We have a system of two particles (1 and 2) with space degrees of freedom obeying a classical relativistic dynamics, plus a quantum internal degree of freedom. In the internal space of each particle we consider an observable Ω corresponding to a Lorentz scalar which can take only the values +1 and -1. The eigenstates of such an observable will be denoted as $|i, + >, |i, - > (i = 1, 2)$

2. We have two macroscopic objects (A and B) - which play the role of measuring apparatuses - again with space degrees of freedom obeying a classical relativistic dynamics and having three possible (internal) states $|X, R>, |X, +>, |X, ->$ $(X = A, B)$. Such objects, which are supposed to be at rest in the reference frame we use to formulate the model, are characterized by a parameter g taking two values $\{0, 1\}$ specifying whether the corresponding apparatus is switched off or on, respectively. The world lines of A and B intersect those of particle 1 and 2 at two space-like points L and R, respectively.

3. The initial conditions are specified by assigning the statevector (in the internal degrees of freedom) of the composite system on a space-like surface σ_0 and by claiming that both apparatuses are in the *ready* state $|X, R>$. In particular we will assume that the initial state is of the singlet type, i.e.,:

$$|\Psi(\sigma_0)> = \frac{1}{\sqrt{2}} \left[|1+, 2-> - |1-, 2+>\right] |A, R>|B, R>. \tag{22}$$

4. We assume completeness of the theory: the assignment of the statevector $|\Psi(\sigma)>$ on any space-like surface σ determines all one can know about the system and no further specification of its state is possible.

5. The dynamics is nonlinear and stochastic and is determined by precise formal rules allowing one to assign a unique statevector to any space-like surface entirely in the future of σ_0. In particular the possible states of the apparatuses A and B determine the outcomes of the measurement processes (local interactions of the systems and the apparatuses). Such rules are quite simple:

   i. Let us consider an arbitrary space like surface σ, (lying entirely in the future of σ_0) and let us denote as $V(\sigma_0, \sigma)$ the hypervolume contained between the two indicated surfaces.

   ii. If neither the space time point L nor R (the points where measurement processes can occur) belong to $V(\sigma_0, \sigma)$, or, if one or both of them belong to it but the corresponding apparatus is switched off, the theory assigns to σ the initial state $|\Psi(\sigma_0)>$.

   iii. Suppose only one of the above points (say L) belongs to $V(\sigma_0, \sigma)$ and the corresponding apparatus is on. Then the statevetor $|\Psi(\sigma)>$ is either $|1+, 2->|A, +>|B, R>$ or $|1-, 2+>|A, ->|B, R>$, the two instances occurring at random, with equal probabilities.

   iv. If consideration is given to a further surface $\hat{\sigma}$ in the future of σ, the evolution from σ to $\hat{\sigma}$ is governed by the same rules, i.e., if R

does not belong to the volume $V(\sigma, \hat{\sigma})$ (or if the apparatus at R is switched off) the state remains the one associated to σ. On the contrary, if $R \in V(\sigma, \hat{\sigma})$ and the apparatus B is on then the only change of the statevector derives from the fact that the state $|B, R >$ becomes $|B, - >$ or $|B, + >$ according to which of the two states under iii is the one associated to σ. In other words the second apparatus simply registers the state of the system triggering it.

We stress that the dynamics is entirely formulated in terms of the covariant statement that certain precise points belong or do not belong to a definite space-time volume. As it is evident, the evolution is genuinely stochastic, it obeys the obvious composition law (one can consistently go from σ_0 to σ and then from σ to $\hat{\sigma}$ or directly from σ_0 to $\hat{\sigma}$ - obviously keeping in mind the various possible stochastic evolutions), it is irreversible, and it guarantees the perfect correlations between the outcomes of the apparata, when they are both on.

Just to give a more precise physical meaning to the process we can think of the internal degrees of freedom of the particles as referring to isospin states, so that the state $|+>$ ($|->$) might, e.g., correspond to a K^+ (K^-) meson.

6.2. Events and Properties in the Relativistic Reduction Model

Let us recall the concept of *event*, by keeping in mind that the appropriate way to define it is the one proposed by A. Shimony. The necessity to resort to such an enlarged meaning of event derives from the fact that the theory has taught us that, if completeness is assumed, then at least at the microlevel indefiniteness of certain properties must be accepted. With reference to the simplified model of the previous subsection we note that nobody would doubt that if no measurement takes place there can be no matter of fact about the particle which is at an objective space-time point being a K^+ or a K^-.

Assuming that the event we are interested in is that of *the definiteness or indefiniteness* of the physical variable Ω_i we can now perform our analysis, i.e., we show how one can consistently introduce covariant criteria concerning events. In particular we investigate what can be said concerning any conceivable space-time point and any possible physical situation.

Let us discuss the following case. We have two particles in the isospin singlet, an apparatus which is on at the space-time point R and we consider a point P on the world line of particle 1 which is space-like with respect to R (obviously a completely analogous argument can be developed for the other particle). We raise the question: what can be said about the event concerning the isospin properties at P of this particle?

Obviously, any statement we can legitimately make must respect the relativistic requirements. For lack of space we will not analyze the problem under consideration in all details (for this purpose see Ghirardi and Grassi 1994; Ghirardi 1996; and Ghirardi 1999). We simply put forward our covariant criterion for property attribution to an individual physical system:

Given the point P on the world line at left, consider the space-like surface σ(P) consisting of the backward light cone originating from P and, outside it, by the space-like surface σ_0 on which the initial conditions are given. The theory assigns a precise statevector to σ(P). Let $\Omega^{(1)}$ be the local property of system 1 we are interested in. If the statevector $|\Psi(\sigma(P))>$ is an eigenstate of $\Omega^{(1)}$ we claim that such property is definite and has the value corresponding to the eigenvalue; if it is not, the property is indefinite.

Some remarks are appropriate:

1. The criterion is unambiguous and manifestly covariant.

2. If one has only one particle, the criterion agrees with the natural request that performing a measurement, and getting an outcome, guarantees that repetition of the same measurement will give the same result.

3. If there is no apparatus on the world line of particle 1, the above criterion implies that the property associated to $\Omega^{(1)}$ is indefinite up to the space-time point at which the future light cone from R intersects the world line of particle 1: the emergence of a property as a consequence of a measurement process occurs luminally.

4. If both apparatuses are on and are space-like separated the property becomes definite as soon as we cross the point L on the world line of system 1. The perfect quantum anticorrelations of the outcomes are reproduced. However, the dynamics and the above prescription are such that one cannot claim that it has been the measurement at R which has caused the outcome at L or vice versa.

5. The general scheme, the dynamics leading to reductions, as well as their physically relevant implications, agree perfectly with the requests that have been put forward by Aharonov and Albert. The theory represents therefore the only known explicit example of a relativistic reduction mechanism exhibiting all necessary features.

7 Nonlocality and Counterfactual Arguments

The situation we have just analyzed can be considered as an explicit and clear example of an EPR-like situation in a genuinely relativistic quantum framework. Therefore, with reference to it, it is particularly appropriate to discuss Einstein's worries about spooky actions at a distance. To this purpose we recall, first of all that Einstein's argument is entirely based on a counterfactual statement. In simple words he claims that if one can predict with certainty the outcome of a possible measurement, then there is an objective property corresponding to the outcome. Rephrasing his position with reference to our example, the measurement at R allows the observer at R to infer the outcome of a possible measurement at L, so that there is an objective property of the system at L and it has been made definite just by the act of measurement at R.

For those who have followed us it should be clear that one should pay a particular attention before drawing such a conclusion. Let us see why it is so.

1. Any counterfactual argument requires a precise specification of the accessibility sphere, and, if one wants to argue counterfactually within a relativistic and nonlocal context, one has to define the accessibility sphere in a way which takes into account both the above features of the theory. In particular, relativity requires to characterize in a covariant way the accessibility sphere.

2. It goes without saying that the natural definition of the accessibility sphere from the actual world should mirror our prescription for property attribution: if consideration is given to a certain space-time point P, the worlds which are nearer, in the counterfactual sense, to the actual world, are those for which the same laws of our world hold and which coincide with the actual world within the past light cone originating from P. We stress that if one takes this attitude, no definite property for particle 1 emerges up to the moment in which the future light cone from R crosses the world line of particle 1. Thus, in the theory there are no spooky actions at a distance.

3. One could be tempted to claim that the appropriate accessibility sphere should consist of all the worlds in which the outcome at R is kept fixed. If this would be the case, then, since the theory guarantees the perfect quantum correlations between the outcomes, one could actually state that the measurement at R has induced the instantaneous emergence of the definite property at L. Such an argument is easily proved to be untenable just because of the unavoidable nonlocal features of the theory. To understand this point one could make reference to a hidden variable theory which determines all conceivable outcomes (such a theory can be easily exhibited for the case under discussion). One could then raise the question: how should one choose the accessibility sphere? And one could suggest that it should consist of all worlds in which the hidden variables have the same value. But if one makes such a choice, one has to keep in mind that, due to the nonlocal nature of the theory, for the

same value of the hidden variables it must unavoidably happen that the outcome at R (or at L) depends on the fact that only one or both apparatuses are switched on. So, this choice does not lead to identify as the accessibility sphere the one made by all worlds in which the outcome at R is kept fixed. We stress that the crucial role of nonlocality is already embodied in our relativistic formalism.

Concluding: the counterfactual argument leading to Einstein's worries about spooky actions at a distance cannot be developed within a correct theoretical framework for the reduction process. To fully clarify this subtle point we urge the reader to consider the two following statements which an observer at R having performed a measurement might make:

i.   I have made a measurement and I have got the result +1. I know that there is an apparatus at L (at a space-like separation from me) which is switched on. Since I know that the final outcomes will be perfectly anti-correlated I can claim that that apparatus will register the outcome -1.

This argument, as the reader has certainly grasped, is perfectly legitimate within relativistic CSL. In fact, as we have discussed in subsection 6.2, if one considers a point $\tilde{R}$ (following R) on the world line of particle 2, the hypersurface $\sigma(\tilde{R})$ consisting of the backward light cone from $\tilde{R}$ and σ_0 is such that $|\Psi(\sigma(\tilde{R}))>$ is either $|1+,2->|AR>|B->$ or $|1-,2+>|AR>|B+>$. The subsequent evolution to a surface which has crossed the point L induces simply the change of $|AR>$ into either $|A+>$ or $|A->$.

ii.   I have made a measurement and I have got the result +1. Since I know that the final outcomes will be perfectly anticorrelated I can claim that if there would be an apparatus which is switched on at L (but in the actual world there is no such apparatus) that apparatus will register the outcome -1. This argument, as shown above, is incorrect because, in confronting the actual world (no apparatus at L) with the accessible one (an apparatus at L which is on), it does not take appropriately into account the crucial role of nonlocality.

Summarizing: reasoning i) is in perfect agreement with the relativistic and nonlocal reduction scheme we have proposed in this paper, but it makes reference to a precise state of affairs, i.e., to the fact that there is an apparatus at L which is switched on. So the argument is by no means a counterfactual one. On the contrary, reasoning ii) is illegitimate within the considered context just because it ignores the crucial role played by nonlocality within a quantum context. It is the combined use of relativistic requirements and of the nonlocal features of the theory which makes unacceptable this kind of counterfactual argument.

Finally, we recall that counterfactual arguments are perfectly possible and legitimate even within our scheme by that they must be developed by resorting to our consistent and covariant prescription for the identification of the accessibility sphere. However, as we have made clear, they do not lead to the conclusion that the theory entails spooky actions at a distance.

References

Aharonov, Y. and Albert, D. Z. (1980) *Physical Review D*, 21, 3316-3324.

Aharonov, Y. and Albert, D. Z. (1981) *Physical Review Letters D*, 24, 359-370.

Aharonov, Y., Albert, D. Z. and Vaidman, L. (1986) *Physical Review D*, 34, 1805-1813.

Bell, J. S. (1987) In C. W. Kilmister (Ed.), *Schrödinger, Centenary Celebration of a Polymath*. Cambridge: Cambridge University Press.

Bohr, N. and Rosenfeld, L. (1933) *Zur Frage der Messbarkeit der elektromanetischen Feldgrössen*, Det. Kgl. Danske Videuskaberues Selskab. Kopehagen: Levin & Munksgaard.

Cohen, O. and Hiley, B. J. (1995) *Foundations of Physics*, 25, 1669-1698.

Ghirardi, G. C. (1996) *Foundations of Physics Letters*, 9, 313-355.

Ghirardi, G. C. (1999) In H. P. Breuer and F. Petruccione (Eds.), *Open Systems and Measurement in Relativistic Quantum Theory*, Lecture Notes in Physics, Springer.

Ghirardi, G. C. and Grassi, R. (1994) *Studies in the History and Philosophy of Science*, 25, 397-423.

Ghirardi, G. C., Grassi, R. and Pearle, P. (1990) *Foundations of Physics*, 20, 1271-1316.

Ghirardi, G. C., Pearle, P. and Rimini, A. (1990) *Physical Review A*, 42, 78-89.

Ghirardi, G. C., Rimini, A. and Weber, T. (1986) *Physical Review D*, 34, 470-491.

Hellwig, K. E. and Kraus, K. (1970) *Physical Review D*, 29, 566-571.

Landau, L. and Peierls, R. (1931/1965) *Z. Phys.* 69, 56. In P. Ter Haar (Ed.) (1965) *Collected papers of L. D. Landau*. Pergammon Press, 40-51.

Pearle, P. (1989) *Physical Review A*, 39, 2277-2289.

Pearle, P. (1990) In A. Miller (Ed.), *Sixty-Two Years of Uncertainty*. New York: Plenum Press.

Shimony, A. (1986) In R. Penrose and C. Isham (Eds.), *Quantum Concepts in Space and Time*. Oxford: Oxford University Press.